MY REVISION NOTES

Cambridge National

Level 1/Level 2

ENGINEERING DESIGN

For the J822 specification

Andy Topliss

T0187382

HODDER EDUCATION
AN HACHETTE UK COMPANY

Photo credits

page 12 (top) © Salazar Benjamin / Shutterstock.com; **page 13 (top)** © RioPatuca / stock. adobe.com; **page 14 (bottom)** © sezer66 / stock.adobe.com; **page 17** © fizkes / stock.adobe. com; **page 20** © simone_n / stock.adobe.com; **page 21** © Oleksandr / stock.adobe.com; **page 23** © tashka2000 / stock.adobe.com; **page 25** © Sashkin / stock.adobe.com; **page 26 (left)** © Yorkshire Pics / Alamy Stock Photo; **page 26 (right)** © Andrew Harker / Shutterstock. com; **page 27 (top)** © Gorodenkoff / stock.adobe.com; **page 29 (bottom)** © Mikalai Bachkou / stock.adobe.com; **page 31 (top)** © Funtay / Stutterstock.com; **page 31 (bottom)** © FOTOGRIN / Shutterstock.com; **page 33 (top)** © Konstantin Z / stock.adobe.com; **page 34** © littlewolf1989 / stock.adobe.com; **page 35** © Gorodenkoff / stock.adobe.com; **page 36 (top)** © Antonio Baccardi / Shutterstock.com; **page 36 (bottom)** © Vitalii / stock.adobe.com; **page 37** © Rawpixel.com / stock.adobe.com; **page 38** © Paul / stock.adobe.com; **page 40** © DenisProduction.com / Shutterstock.com; **page 42** © Alfons Photographer / stock.adobe.com; **page 43 (middle)** © Marzky Ragsac Jr. / stock.adobe.com; **page 65 (top)** © Gorodenkoff / Shutterstock.com; **page 65 (bottom)** © stokkete / stock.adobe.com; **page 66 (top right)** © Peeradontax / Shutterstock.com; **page 66 (bottom)** © Kitti / stock.adobe.com; **page 67** © fotofabrika / stock.adobe.com; **page 69 (steel rule)** © AVD / stock.adobe.com; **page 69 (tape measure)** © Александр Текучев / stock.adobe.com; **page 69 (vernier calliper)** © sergojpg / stock.adobe.com; **page 69 (micrometer)** © Andrzej Tokarski / stock.adobe.com; **page 69 (weight scales)** © injenerker / stock.adobe.com; **page 69 (multimeter)** © krasyuk / stock.adobe.com; **page 70** © weedezign / stock.adobe.com

Although every effort has been made to ensure that website addresses are correct at time of going to press, Hodder Education cannot be held responsible for the content of any website mentioned in this book. It is sometimes possible to find a relocated web page by typing in the address of the home page for a website in the URL window of your browser.

Hachette UK's policy is to use papers that are natural, renewable and recyclable products and made from wood grown in well-managed forests and other controlled sources. The logging and manufacturing processes are expected to conform to the environmental regulations of the country of origin.

Orders: please contact Hachette UK Distribution, Hely Hutchinson Centre, Milton Road, Didcot, Oxfordshire, OX11 7HH. Telephone: +44 (0)1235 827827. Email: education@hachette. co.uk. Lines are open from 9 a.m. to 5 p.m., Monday to Friday. You can also order through our website: www.hoddereducation.co.uk

ISBN: 978 1 3983 5247 6

© Andy Topliss, 2022

First published in 2022 by
Hodder Education,
An Hachette UK Company
Carmelite House
50 Victoria Embankment
London EC4Y 0DZ

www.hoddereducation.co.uk

Impression number   10 9 8 7 6 5 4 3 2

Year   2026 2025 2024

Cover photo © Aranami – stock.adobe.com

Typeset in India

Printed in Spain

A catalogue record for this title is available from the British Library.

Get the most from this book

This book will help you revise for your Cambridge National in Engineering Design exam (Unit R038: Principles of engineering design). You can find out more about the exam on pages 7–9.

Everyone has to decide his or her own revision strategy, but it is essential to review your work, learn it and test your understanding. These Revision Notes will help you to do that in a planned way, topic by topic. Use this book as the cornerstone of your revision and don't hesitate to write in it: personalise your notes and check your progress by ticking off each section as you revise.

Tick to track your progress

Use the revision planner on pages 4 and 5 to plan your revision, topic by topic. Tick each box when you have:

+ revised and understood a topic
+ tested yourself
+ practised exam questions.

You can also keep track of your revision by ticking off each topic heading in the book. You may find it helpful to add your own notes as you work through each topic.

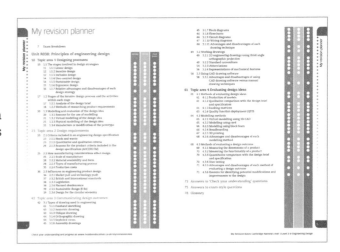

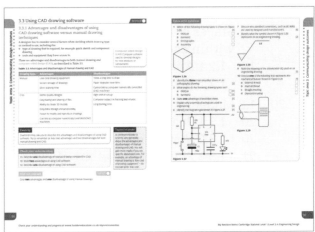

Features to help you succeed

tips

tips are given throughout the book to help with your exam technique and maximise your the exam.

stakes

entifies common mistakes made by and offers guidance on how to avoid

rstanding

ns test your knowledge and u work through the course. the back of the book.

Now test yourself

These revision activities will guide your note-taking.

Definitions and key words

Clear, concise definitions of essential-to-know terms are provided where they first appear.

Exam-style questions

Practice exam questions are provided for each topic. Use them to consolidate your revision and practise your exam skills. Answers are given at the back of the book.

My revision planner

7   Exam Breakdown

Unit R038: Principles of engineering design

10   Topic area 1: Designing processes
  10   1.1 The stages involved in design strategies
    11   1.1.1 Linear design
    12   1.1.2 Iterative design
    12   1.1.3 Inclusive design
    14   1.1.4 User-centred design
    14   1.1.5 Sustainable design
    15   1.1.6 Ergonomic design
    16   1.1.7 Relative advantages and disadvantages of each
          design strategy
  17   1.2 Stages of the iterative design process and the activities
        within each stage
    17   1.2.1 Analysis of the design brief
    18   1.2.2 Methods of researching product requirements
  20   1.3 Modelling and evaluation of the design idea
    20   1.3.1 Reasons for the use of modelling
    20   1.3.2 Virtual modelling of the design idea
    21   1.3.3 Physical modelling of the design idea
    21   1.3.4 Manufacture or modification of the prototype

23   Topic area 2: Design requirements
  23   2.1 Criteria included in an engineering design specification
    23   2.1.1 Needs and wants
    23   2.1.2 Quantitative and qualitative criteria
    24   2.1.3 Reasons for the product criteria included in the
          design specification (ACCESS FM)
  25   2.2 How manufacturing considerations affect design
    25   2.2.1 Scale of manufacture
    28   2.2.2 Material availability and form
    28   2.2.3 Types of manufacturing process
    35   2.2.4 Production costs
  36   2.3 Influences on engineering product design
    36   2.3.1 Market pull and technology push
    37   2.3.2 British and international standards
    38   2.3.3 Legislation
    38   2.3.4 Planned obsolescence
    39   2.3.5 Sustainable design (6 Rs)
    39   2.3.6 Design for the circular economy

42   Topic area 3: Communicating design outcomes
  42   3.1 Types of drawing used in engineering
    42   3.1.1 Freehand sketching
    43   3.1.2 Isometric drawing
    43   3.1.3 Oblique drawing
    44   3.1.4 Orthographic drawing
    44   3.1.5 Exploded views
    45   3.1.6 Assembly drawings

REVISED   TESTED   EXAM READY

Check your understanding and progress at **www.hoddereducation.co.uk/myrevisionnotes**

45 3.1.7 Block diagrams
46 3.1.8 Flowcharts
46 3.1.9 Circuit diagrams
47 3.1.10 Wiring diagrams
48 3.1.11 Advantages and disadvantages of each
       drawing technique
49 3.2 Working drawings
49 3.2.1 2D engineering drawings using third angle
       orthographic projection
49 3.2.2 Standard conventions
54 3.2.3 Abbreviations
56 3.2.4 Representations of mechanical features
58 3.3 Using CAD drawing software
58 3.3.1 Advantages and disadvantages of using
       CAD drawing software versus manual
       drawing techniques

61 **Topic area 4: Evaluating design ideas**
61 4.1 Methods of evaluating design ideas
61 4.1.1 Production of models
61 4.1.2 Qualitative comparison with the design brief
       and specification
61 4.1.3 Ranking matrices
63 4.1.4 Quality function deployment (QFD)
64 4.2 Modelling methods
64 4.2.1 Virtual modelling using 3D CAD
65 4.2.2 Modelling using card
66 4.2.3 Modelling using block foam
66 4.2.4 Breadboarding
67 4.2.5 3D printing
68 4.2.6 Advantages and disadvantages of each
       modelling method
68 4.3 Methods of evaluating a design outcome
69 4.3.1 Measuring the dimensions of a product
70 4.3.2 Measuring the functionality of a product
70 4.3.3 Quantitative comparison with the design brief
       and specification
70 4.3.4 User testing
70 4.3.5 Advantages and disadvantages of each method of
       evaluating a design outcome
71 4.3.6 Reasons for identifying potential modifications and
       improvements to the design

73 Answers to 'Check your understanding' questions

75 Answers to exam-style questions

78 Glossary

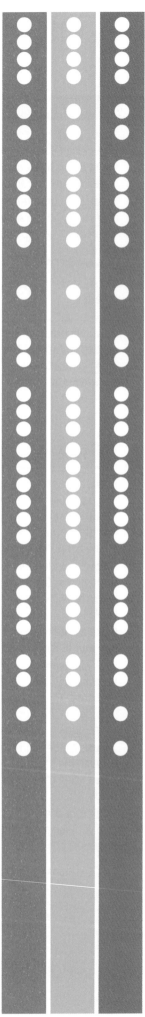

My revision planner

Countdown to my exam

6–8 weeks to go

+ Start by looking at the specification — make sure you know exactly what material you need to revise and the style of the exam. Use the revision planner on pages 4 and 5 to familiarise yourself with the topics.

+ Organise your notes, making sure you have covered everything on the specification. The revision planner will help you group your notes into topics.

+ Work out a realistic revision plan that will allow you time for relaxation. Set aside days and times for all the subjects that you need to study, and stick to your timetable.

+ Set yourself sensible targets. Break your revision down into focused sessions of around 40 minutes, divided by breaks. These Revision Notes organise the basic facts into short, memorable sections to make revising easier.

REVISED ◯

4–6 weeks to go

+ Read through the relevant sections of this book and pay close attention to the exam tips, typical mistakes and key terms. Tick off the topics as you feel confident about them. Highlight those topics you find difficult and look at them again in detail.

+ Test your understanding of each topic by working through the 'Check your understanding' questions in this book. Look up the answers at the back of the book.

+ Make a note of any problem areas as you revise, and ask your teacher to go over these in class.

+ Look at past papers. They are one of the best ways to revise and practise your exam skills. Write or prepare planned answers to the exam-style questions provided in this book. Check your answers at the back of the book.

+ Try using different revision methods as you work through the sections. For example, you can make notes using mind maps, spider diagrams or flash cards.

+ Track your progress using the revision planner and give yourself a reward when you have achieved your target.

REVISED ◯

One week to go

+ Try to fit in at least one more timed practice of an entire past paper and seek feedback from your teacher, comparing your work closely with the mark scheme.

+ Check the revision planner to make sure you haven't missed out any topics. Brush up on any areas of difficulty by talking them over with a friend or getting help from your teacher.

+ Attend any revision classes put on by your teacher. Remember, your teacher is an expert at preparing people for exams.

REVISED

The day before the exam

+ Flick through these Revision Notes for useful reminders, for example the exam tips, typical mistakes, key terms and exam checklists.

+ Check the time and place of your exam.

+ Make sure you have everything you need — extra pens and pencils, tissues, a watch, bottled water, sweets.

+ Allow some time to relax and have an early night to ensure you are fresh and alert for the exam.

REVISED

My exams

Unit R038 paper

Date: ....

Time: ....

Location: ....

Exam breakdown

About the exam

Unit R038 of the Cambridge National Level 1/2 in Engineering Design is about developing knowledge, understanding and practical skills that would be used in the engineering design and development sector. This is a compulsory examined unit with a one-hour, 15-minutes paper, which is worth 70 marks.

Question types

On your exam paper, there will be a range of different question types, such as:
+ multiple choice
+ completion of tables
+ completion of drawings
+ extended-answer questions.

Some questions are a mixture of these question types, ranging from 1 to 4 marks.

Some extended-answer questions may be worth 6 marks. These are assessed against a 'levels' mark scheme. These levels relate to the written quality of your answer. You should write in a structured way with accurate spelling, punctuation and grammar, and use specialist technical terms where you can, to achieve the higher levels.

Each level has a list of required content to achieve that level.

Example of levels of response descriptions	
Level 3 (5–6 marks)	A detailed response that: + shows detailed knowledge and understanding + makes many points, most of which are well developed + is well structured and consistently uses appropriate technical terms + has few, if any, errors in grammar, punctuation and spelling.
Level 2 (3–4 marks)	An adequate response that: + shows good knowledge and understanding + makes some valid points, a few of which may be developed + is reasonably well structured and uses some appropriate technical terms + has occasional errors in grammar, punctuation and spelling.
Level 1 (1–2 marks)	A basic response that: + shows limited knowledge and understanding + makes some basic points, which are rarely developed + has limited coherence and structure, with little or no use of appropriate technical terms + has errors in grammar, punctuation and spelling, which may be noticeable and intrusive.
0 marks = nil response or no response worthy of credit	

Command words

When sitting your exam, read each question carefully and identify exactly what is required. You might want to highlight or underline any key words that you think will help you understand what the question is asking for. If you do this, always highlight the command word, as this will help you to plan the content of your answer.

The following table contains the command words that could be used at the start of questions in your exam and explains what each requires you to do.

Command words	What you must do
Analyse	Separate or break down information into parts and identify their characteristics or elements.
	Explain the pros and cons of a topic or argument and make reasoned comments.
	Explain the impacts of actions using a logical chain of reasoning.
Annotate	Add information, for example to a table, diagram or graph until it is final.
	Add all the required or appropriate parts.
Calculate	Provide a numerical answer, showing how it has been worked out.
Choose	Select an answer from the options given.
Circle	Select an answer from the options given.
Compare and contrast	Give an account of the similarities and differences between two or more items or situations.
Complete	Add information, for example to a table, diagram or graph until it is final.
	Add all the required or appropriate parts.
Create	Produce a visual solution to a problem (for example a mind map, flowchart or visualisation).
Describe	Give an account, including all the relevant characteristics, qualities or events.
	Give a detailed account of something.
Discuss	Present, analyse and evaluate relevant points (for example for/against an argument).
Draw	Produce a picture or diagram.
Evaluate	Make a reasoned qualitative judgement considering different factors and using available knowledge and/or experience.
Explain	Give reasons for and/or causes of something.
	Use words or phrases such as 'because', 'therefore' or 'this means that' in answers.
Fill in	Add information, for example to a table, diagram or graph until it is final.
	Add all the required or appropriate parts.
Identify	Select an answer from the options given.
	Recognise, name or provide factors or features.
Justify	Give good reasons for offering an opinion or reaching a conclusion.
Label	Add information, for example to a table, diagram or graph until it is final.
	Add all the required or appropriate parts.
Outline	Provide a short account, summary or description.
State	Give factors or features.
	Provide short, factual answers.

Key points to remember in the exam

REVISED

+ When writing your answer, produce a response that is clear and concise. Try not to waffle.
+ Make sure you do not repeat information that is already given in the wording of the question.
+ If a question wants you to apply your knowledge and understanding, you need to use examples.
+ Look at how many parts there are to a question and make sure you answer all of them.
+ Check how many marks your question is worth and match your answers to the number of marks in the question. Mark allocations are provided in square brackets [ ] at the end of each question or part question.
+ Try not to miss out any questions. You could pick up a mark with an educated guess.
+ You don't have to answer the questions in order. If you don't know the answer straight away, don't spend time being stuck – move on to a question you can do and come back later.

Check your understanding and progress at **www.hoddereducation.co.uk/myrevisionnotes**

- Write your answers clearly in the spaces provided in the answer booklet.
- If you need more space to complete an answer, use the additional lined pages at the end of the answer booklet and clearly number the question(s) where additional answers have been written.
- Avoid writing anything you want to be marked in the margins and always indicate if you run out of space that your answer continues on additional paper or at the end of the answer booklet.
- The examiner needs to be able to read your answer, so keep your handwriting neat.
- The quality of written communication (QWC) is assessed via the final extended question, so focus on making this answer your best in terms of how you structure and write it.

Revising for your exam

There are lots of different ways to revise for your exam, and you may find some revision methods work better than others. Here are some ideas to help you:

- Mind maps: read through a topic and then, without your notes, put the key points into a mind map. Check to see if you have covered everything, and if not, add the missing knowledge to the mind map. You will then have a concise version of your topic notes. A mind map for drawing types used in engineering could look like Figure A.
- Exam questions: as well as completing the exam-style questions in this book, you can also visit the OCR website where there are lots of past papers and mark schemes. You can use these to test your knowledge. You will also become familiar with the types of questions that could appear on your paper. Try to answer a whole paper in one hour without stopping, so that you get used to the amount of time you have available. This will prepare you for exam conditions.
- Revision cards: simply read a topic and rewrite your notes briefly on small cards. Make sure you include all the main points. You may also wish to put notes on one side of the card and questions and answers on the other.
- Study buddy: revise with a friend and test one another.

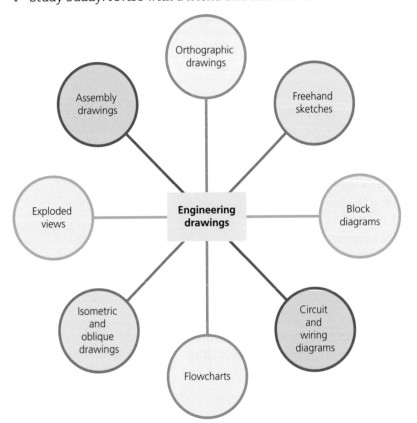

Figure A Example of a mind map

My Revision Notes Cambridge National Level 1/Level 2 in Engineering Design

Topic area 1: Designing processes

Ideas do not just appear from thin air but are the result of trying to solve a problem, for example trying to improve something or make it easier to use. Various design processes have been developed to focus on different needs and wants.

In this chapter, you will learn about the design processes used to design products.

1.1 The stages involved in design strategies

REVISED ●

Design strategies are a series of steps used to plan a design process. They offer useful ways to think of and manage ideas to create solutions.

There are six different types of design strategy, with each focusing on problems in a different way (see Figure 1.1).

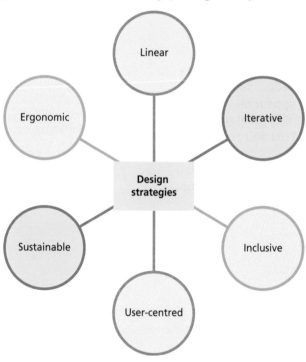

Figure 1.1 Types of design strategy

Different design strategies will be used for different situations, based on the problem to be solved and the specific needs and wants of the users. The choice of design strategy will depend on many factors, including the following:

+ The type of product being designed, for example:
  + Will it be mass produced to be used by everyone or used by a limited section of people?
  + Does it need to be recycled?
+ Product life cycle, for example:
  + How will the product evolve?
  + What will the product's effect be on the world from its first appearance, in the way it is manufactured and used, to what happens to it when it is no longer used?

> **Users** People who will use the final product
>
> **Product life cycle** The various stages of a product's evolution, from its beginning to its end

10

- Human factors – most products are designed for people, so human factors such as hearing, vision, manual dexterity, strength and reach may need to be taken into consideration.
- Cost – there is a cost for the time people take to design a product, fully test it and prototype a new idea. Other costs of production include materials, manufacturing, labour and factory overheads.

> **Exam tip**
>
> Questions on this topic require you to know the definitions of each type of design strategy and their relative advantages and disadvantages.

Topic area 1: Designing processes

Manual dexterity The skill of using the hands to carry out a task with precision

Prototype To create a 3D model that demonstrates the functionality of a product

Overheads Costs or expenses, such as lighting, heating and equipment, that are paid out by an organisation

Linear design Development of a product through a series of sequential stages

1.1.1 Linear design

Linear design uses strict controls to manage risk in the design process. Each phase must be fully completed before going on to the next. This allows designers to catch any errors earlier on, when they are least expensive and time consuming to fix. This method is used in traditional engineering organisations, where the design moves from one stage to the next. Each stage typically represents a department or an area within an organisation.

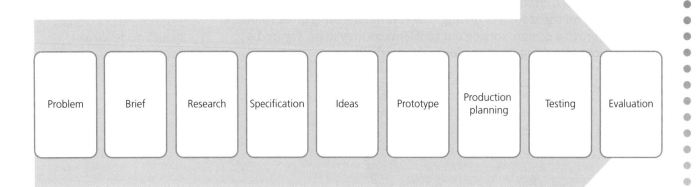

Figure 1.2 Linear design

For example:
- The design department of a steel beam manufacturer will spend a lot of time planning and researching the design and working out the beam's structural properties. It will then design the beam using the ideas from the planning stage.
- The design will then go to the drawing department, where engineering drawings are produced.
- These engineering drawings will pass to the manufacturing department, in order to create a prototype.
- The testing department will test the prototype, and if it functions correctly the final design will pass to the manufacturing production department.

As the design has to go from department to department, it makes it difficult to go back and fix earlier problems.

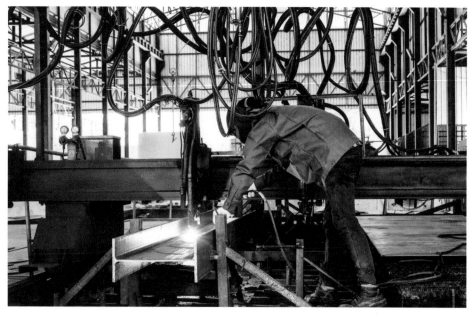

Figure 1.3 Steel beam manufacturers follow a linear design process

1.1.2 Iterative design

Iterative design is a circular process that models, evaluates and develops designs based on the results of testing. The continual testing gives feedback that can improve a design, sorting out problems as they arise. Figure 1.4 shows the iterative design process for making a drill.

Iterative design
Development of a product through modelling and repeated testing

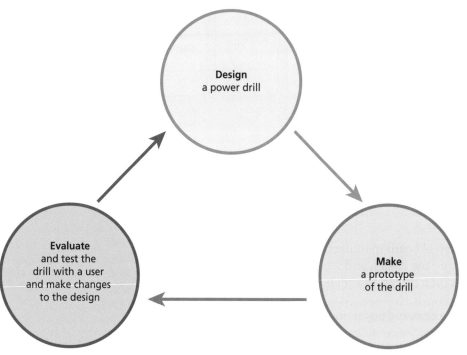

Figure 1.4 Iterative design process for making a drill

1.1.3 Inclusive design

Inclusive design focuses on developing products that can be used by as many people as possible, regardless of their age, ability or background. It aims to enable everyone to participate equally and independently in everyday activities. For example, architects design buildings that allow access by all users, by taking into consideration the different needs of certain individuals, such as children, older people and disabled people.

Inclusive design
Development of a product so that it can be used by as many people as possible, regardless of their age, ability or background

Check your understanding and progress at **www.hoddereducation.co.uk/myrevisionnotes**

Figure 1.5 This ramp allowing access by disabled users is an example of inclusive design

Figure 1.6 summarises the five focal points for inclusive design and gives an example of each.

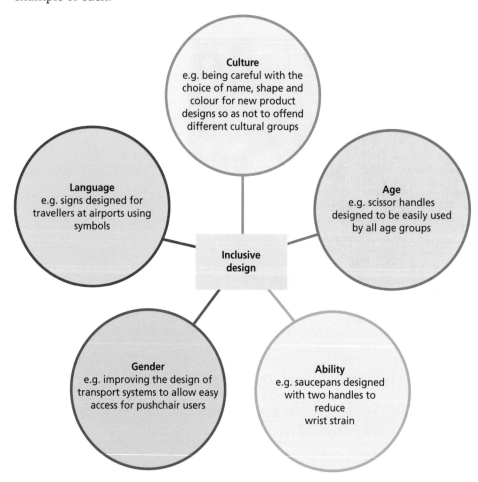

Culture
e.g. being careful with the choice of name, shape and colour for new product designs so as not to offend different cultural groups

Age
e.g. scissor handles designed to be easily used by all age groups

Language
e.g. signs designed for travellers at airports using symbols

Inclusive design

Gender
e.g. improving the design of transport systems to allow easy access for pushchair users

Ability
e.g. saucepans designed with two handles to reduce wrist strain

Figure 1.6 Inclusive design

1.1.4 User-centred design

User-centred design is based on a clear understanding of user needs and requirements, or a target market. Time is spent gathering user feedback during the early design stages of a product. This strategy is not only used for physical products but also for phone apps, computer programmes and gaming, where the user may be heavily involved in the development of the product.

User-centred design
Development of a product with a clear understanding of user needs and requirements

Target market The group of people a product is made for

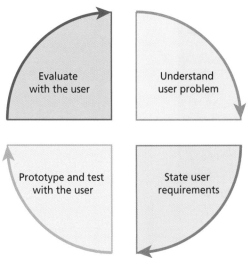

Evaluate with the user

Understand user problem

Prototype and test with the user

State user requirements

Figure 1.7 User-centred design

Figure 1.8 User-centred design is often used for gaming

1.1.5 Sustainable design

Sustainable design seeks to avoid harm to people or the planet. During the research stages of the design process, designers will look at how a product might affect the environment through its life cycle (Figure 1.9) and try to work with the 6 Rs of sustainability (see Table 1.1 and Section 2.3.5).

Sustainable design
Development of a product while trying to reduce negative impacts on the environment

Check your understanding and progress at **www.hoddereducation.co.uk/myrevisionnotes**

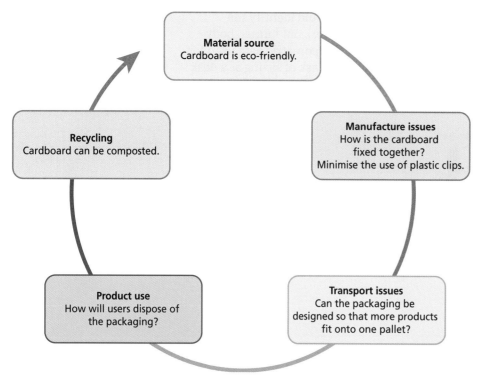

Figure 1.9 Sustainable design and product life cycle for cardboard packaging

Table 1.1 The 6 Rs of sustainability

Reduce	Cut down on the amount of materials used.
Reuse	Use a product to make something else.
Recycle	Reprocess a product to make something else.
Rethink	Do we need to make so many products?
Refuse	Don't buy a product if you don't need it.
Repair	When a product breaks down, try to fix it.

1.1.6 Ergonomic design

Ergonomic design takes into consideration how people interact with the products they use and the environments they use them in. It uses anthropometric data to ensure products are comfortable to use, are easy to understand and fit the user they are designed for. Therefore, a designer needs to understand the different variations of the human body and human preferences. Figure 1.10 shows an example of ergonomic design in the workplace.

> **Ergonomic design**
> Development of products using anthropometric data so that they perfectly fit the people who use them
>
> **Anthropometric**
> Relating to the study of measurements of the human body

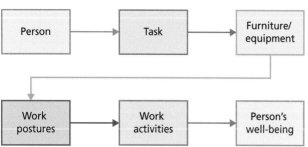

Figure 1.10 Ergonomic design in the workplace

When designing an office computer chair, a designer would identify the average-sized person using anthropometric tables to work out human body measurements. Figure 1.11 shows an example of anthropometric data used for computer users. The designer would then consider the task of sitting at a desk and operating a computer for many hours and how this affects the person's posture and well-being.

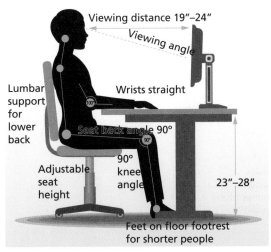

Viewing distance 19"–24"
Viewing angle
Lumbar support for lower back
Wrists straight
Seat back angle 90°
90° knee angle
Adjustable seat height
23"–28"
Feet on floor footrest for shorter people

Figure 1.11 Example of anthropometric data for computer users

1.1.7 Relative advantages and disadvantages of each design strategy

Table 1.2 Advantages and disadvantages of design strategies

Design strategy	Advantages	Disadvantages
Linear	Manages risk Catches errors early in the design process	Offers no flexibility to make changes in order to improve a design
Iterative	Evaluates and improves designs Allows continual prototyping and testing	Expensive due to continual prototyping and testing
Inclusive	Focuses on users with specific needs Enables a wide range of users to participate in everyday activities	Cannot cater for all users Inclusive products are more expensive than generic products
User-centred	Allows a clear understanding of user needs via user feedback Involves the user in the design process	Relies on good user feedback, which may be limited depending on who is chosen as a user
Sustainable	Ensures products cause minimal harm to people and the planet Uses the 6 Rs of sustainability	Requires the attitudes of people and industry to change for it to work
Ergonomic	Takes into consideration variations in the human body	Ergonomic products are more expensive than generic products

Check your understanding

1 What is meant by the term product life cycle?
2 Outline the purpose of user-centred design.
3 Identify **two** of the 6 Rs of sustainability.
4 Why do designers use anthropometric data?
5 Identify **one** disadvantage of user-centred design.
6 Identify **one** advantage of sustainable design.

Now test yourself TESTED

Laptops are lightweight, small, easily portable and have battery backup. They allow us to work and game flexibly. However, they have been blamed for causing back, neck and shoulder problems. Understanding this would allow a designer to improve furniture design.

Figure 1.12 Laptops have been blamed for causing back, neck and shoulder problems

1   Identify the adaptations a furniture designer could make to ensure a laptop is safer and more comfortable to use.

2   People use laptops on a regular basis. What are some of the key product features that have been added to laptop design based on the needs of the user?

1.2 Stages of the iterative design process and the activities within each stage

REVISED

Designers use iterative design to improve the design brief of a product, to ensure it meets the specification and the client's requirements. It is a cyclic process of design, make and evaluate until the final product design can be agreed.

The aim of the design process is to help the designer produce a specification based on the design brief and then generate ideas by sketching and modelling.

1.2.1 Analysis of the design brief

A design brief sets out what is required by a client (user) and explains what needs to be done and within what time limits. It contains relevant details but not the solution to the problem.

Design brief The context for a design problem, with reference to user needs and wants, product performance, end use, scale of production, time limits and target market

Specification A list of criteria that a product needs to address

The design brief typically explains:
+ what product needs to be designed
+ what the product will be used for
+ whom the product is for
+ where the product will be sold
+ where the product will be used
+ when the prototype must be finished.

For example, a design brief for a sports bottle may read:

> Design a sports bottle to be used by a marathon runner. It should be lightweight and easily carried while training. It should fit safely in either hand and be able to contain 500 ml of liquid. It will be sold in specialist sports shops and a prototype needs to be prepared within three months.

Once a designer and client have agreed on a design brief, the designer will plan the research and analyse each aspect of the client's requirements. It is important that the design brief is correct the first time, as the designer could easily begin to develop a product that is not fit for purpose.

1.2.2 Methods of researching product requirements

Primary and secondary research

In order to identify and analyse the needs of potential customers, designers will carry out market research. This research will be either primary or secondary:
+ Primary research involves gathering original information first hand.
+ Secondary research is where information is gathered from sources that already exist.

Table 1.3 gives examples of primary and secondary research, and information about their advantages and disadvantages.

Table 1.3 Primary and secondary research

	Primary research	Secondary research
Method	Original information is gathered first hand.	Information is gathered from sources that already exist.
Sources of information	+ Focus groups + Interviews + Observations + Product analysis + Surveys + Industrial visits	+ Internet + TV programmes + Databases + Textbooks + Newspapers
Advantages	The data collected is original and specific. Targeted customers can be asked their opinion.	The data is extensive. It is time and cost effective.
Disadvantages	Collecting the data is time consuming and expensive. Targeted customers may not be appropriate to give a range of opinions.	Data may be inaccurate. Data may not be specific.

Designers, on behalf of their clients, will use both primary and secondary research to focus on the needs of target customers when considering the features of a new product. For example:
+ Primary research, such as focus groups, involves potential customers who may benefit from a new product. They would be shown prototypes to review and asked their opinions.
+ Secondary research, such as sourcing information from the internet, can help designers find existing products to see what is currently selling well.

Market research Process of gathering information about the needs and preferences of potential customers

Interviews with potential users and focus groups

To understand up-to-date opinions and attitudes, designers can carry out personal interviews with one potential user, as well as focus groups with a larger number of potential users:

+ Interviews allow the designer to ask direct questions about a new product and receive in-depth answers. However, there is always a possibility that the interviewee may say what they think the designer wants to hear, rather than giving their own opinion.
+ Focus groups allow the participants to build on each other's answers, and they are often happy to participate in discussion about a new product. However, the groups must be selected carefully, as dominant people may stop others from talking and some individuals may not be confident expressing their opinions in public.

Market research to determine existing products

Designers often use market research to find out what products exist on the market, to get an idea of how successful the products are and what features could be improved. For example, there are a number of companies that sell computer tablets with similar features.

Use of anthropometric data tables

Anthropometric data tables help designers to consider a range of users of a new product. They include measurements for dimensions such as height, hand size, finger length, waist size, reach and grip.

Anthropometric data is grouped into percentiles, which include the 5th percentile (the smallest five per cent of people), the 50th percentile (the average size of people) and the 95th percentile (the largest five per cent of people).

Analysis of existing products using ACCESS FM

ACCESS FM is a useful method for analysing existing products as part of product research. It is an easy acronym to remember, and it covers the majority of the key design criteria:

+ Aesthetics
+ Cost
+ Customer
+ Environment
+ Size
+ Safety
+ Function
+ Materials.

Using these headings, and a series of questions, it is possible to research products systematically and to compare relative strengths and weaknesses across a range of similar products.

Analysis of existing products using product disassembly

Product disassembly, also referred to as reverse engineering, is probably the most useful research designers can undertake. It involves looking carefully at a product and taking it apart to find out how it works and how it was made. It allows designers to look closely at how related products are manufactured and put together.

Production of an engineering design specification

On completion of all the relevant research, a designer will create an engineering design specification using the key criteria of ACCESS FM as a guide. This is an important document, as it details the new product requirements and is agreed with the client.

Aesthetics How a product appeals to the senses; something that is pleasing in appearance based on its form, shape, symmetry, texture, colour and proportion

Disassembly Taking something apart, for example a product or piece of equipment

Engineering design specification Detailed document that defines the criteria required for a new product

> **Exam tip**
>
> Make sure you know what each letter means in ACCESS FM.

19

Generation of design ideas by sketching and modelling

With the engineering design specification agreed, a designer can then generate product design ideas for the client. Generation of design ideas may refer to the creation of the initial design or to the modification/improvement of the existing design.

A wide range of methods may be used, for example:
+ freehand sketching for initial ideas
+ 3D modelling using virtual and physical techniques.

Check your understanding

7 Name **two** types of information that may be included in a design brief.

8 Describe what is meant by primary research.

9 Give **one** meaning of C in ACCESS FM.

10 Give **one** meaning of S in ACCESS FM.

1.3 Modelling and evaluation of the design idea

REVISED ●

Designers will make lots of models as part of the iterative design process. Ideas that are sketched on paper or produced virtually need to be modelled and evaluated to check they fulfil the design brief and specification. To help improve a design, evaluation may be carried out using interviews to get user feedback. This feedback can then be used to inform and make another prototype.

1.3.1 Reasons for the use of modelling

Designers use modelling to check that the proportions, scale and functionality of a product are correct:
+ They can test proportions by checking that the relationship between the size of different parts of a product is correct or attractive.
+ They can test scale by checking that the overall dimensions of the product are correct or attractive.
+ They can test functionality by checking that the product works or operates in a proper or particular way.

1.3.2 Virtual modelling of the design idea

Three-dimensional (3D) models of design ideas can be created virtually using computer-aided design (CAD) software. Using this software, objects are drawn in 2D then converted to 3D, in order to achieve better visualisation of a product and all its components. The function can also be observed, as the components can be animated.

Models Virtual or physical 3D objects that demonstrate the aesthetics of products

Computer-aided design (CAD) Using computer software to develop designs for new products or components

---

Typical mistake

Designers use modelling to check three aspects of a model, the proportion, scale and functionality. A typical mistake is to just refer to the size only.

Figure 1.13 Developing designs for new products and components using CAD

Check your understanding and progress at **www.hoddereducation.co.uk/myrevisionnotes**

Table 1.4 Advantages and disadvantages of virtual modelling

Advantages	Disadvantages
No need to make a physical model	High initial set-up costs
Very accurate modelling	Requires skill training

1.3.3 Physical modelling of the design idea

Models can be made physically using a variety of materials in the workshop. Early models may be made from cheap materials, such as cardboard, to allow quick visualisation of ideas. Later models will be made of more expensive materials, such as metals and plastics.

Table 1.5 Advantages and disadvantages of physical modelling

Advantages	Disadvantages
Allows designers to explore and test product ideas	Can be time consuming
	A level of skill is required
Allows potential users to see and handle a product and give valuable feedback	Can be costly
	Model might not look like the final product

Figure 1.14 Making physical models

1.3.4 Manufacture or modification of the prototype

Before a new product is manufactured, the virtual or physical prototype needs to be compared against the design brief and engineering design specification and approved by the client. If at this stage there are still issues to be corrected, new models will be made until designer and client agreement is achieved.

Check your understanding

11 What is a prototype used for?

12 What is **one** advantage of using 3D CAD?

13 What is **one** disadvantage of making physical models?

Exam tip

Remember the difference between a model and a prototype: a model is either a virtual or physical shape based on a product, while a prototype is a model that can test the function of a design.

Exam-style questions

1 Identify **one** of the following as an example of ergonomic design strategy. [1]
   a Computer-gaming chair ☐
   b Solar-powered lamp ☐
   c Storage container ☐
   d Wheelchair access at a bus stop ☐
   e Doorbell camera ☐

2 State **one** advantage of inclusive design. [1]

3 State **one** disadvantage of sustainable design. [1]

4 State **one** meaning of A in ACCESS FM. [1]

5 Describe, using **one** example, what is meant by secondary research. [2]

6 Identify the **three** stages of iterative design. [3]

7 Describe **two** reasons why modelling is used. [4]

8 Explain, using **one** example, why sustainable design has become more important. [4]

9 Discuss the advantages and disadvantages of using the linear design strategy when designing a lifejacket for a water sports activity. [6]

10 Discuss the advantages and disadvantages of using the sustainable design strategy when designing an electric car. [6]

Exam checklist

In this topic, you learned about the following:
+ The stages involved in design strategies, how each strategy might be applied and their relative advantages and disadvantages:
  + Linear design, which uses a series of sequential stages and strict controls to manage risk in the design process
  + Iterative design, which is a circular process that models, evaluates and improves designs based on the results of testing
  + Inclusive design, which focuses on designs for specific users with specific needs
  + User-centred design, which has a clear understanding of user needs and requirements
  + Sustainable design, which tries to reduce negative impacts on the environment
  + Ergonomic design, which uses anthropometric data to help products fit users perfectly
+ Stages of the iterative design process and the activities within each stage:
  + Analysis of the design brief, which sets out what is required by a client (user) and explains what needs to be done and within what time limits

+ Research methods and how the information obtained using each method contributes to the design process
+ The difference between primary and secondary research and the relative advantages and disadvantages of these when deciding product requirements
+ Use of anthropometric data
+ Use of ACCESS FM as an analysis tool
+ Use of product disassembly as an analysis tool
+ Production of the engineering design specification
+ Generation of design ideas by sketching and modelling, referring to either the creation of the initial design or the modification/improvement of an existing design
+ Modelling and evaluation of the design idea:
  + Reasons why modelling is used: to test proportions, scale and function
  + Virtual modelling of the design idea
  + Physical modelling of the design idea
  + Manufacture and modification of the prototype
  + Comparison of the model or prototype against the requirements of the design brief and specification.

Topic area 2: Design requirements

2.1 Criteria included in an engineering design specification

An engineering design specification is a list of requirements for an engineered product. Designers use this list to advise them what characteristics and features to include.

2.1.1 Needs and wants

A need is a characteristic that a product must have – that is, a basic requirement that makes it fit for purpose. For example, a computer mouse must have finger controls and communicate with a computer.

> **Need** Critical aspect of a product that makes it fit for purpose
>
> **Want** Non-essential but desirable aspect of a product
>
> **Quantitative** Can be expressed as a number and quantified by hard facts
>
> **Objective** Based on facts and reliable; not influenced by personal feelings or opinions
>
> **Qualitative** Cannot be expressed as a number and describes opinions, qualities and feelings
>
> **Subjective** Based on personal feelings, tastes or opinions

Figure 2.1 What are the needs and wants for a computer mouse?

A want is a characteristic of a product that is desirable but not essential and allows for personal preference. Wants may help a product to be more successful, but they are not required for the product to function. For example, a computer mouse could be any colour. This is a want, because the mouse could still be operated no matter what colour it is.

2.1.2 Quantitative and qualitative criteria

Quantitative criteria can be measured. For example, for a computer mouse to fit into an average adult hand, it should be a width of between 60 and 100 mm. It is an advantage of quantitative criteria that objective measurements can be made. This means the measurement is against a known and accepted standard, such as weight in grams and width in millimetres.

Qualitative criteria are based on the opinion of an assessor and therefore subjective. For example, the computer mouse must be easy to operate, and each assessor would either agree or disagree depending on their own experience. Market research is often used when assessing large quantities of products; many users are asked their opinions to allow a majority design decision to be made.

> **Exam tip**
>
> Make sure you know the difference between needs and wants. Needs are characteristics that a product *must* have, and wants are characteristics that are *desirable* in a product.

> **Exam tip**
>
> Make sure you know the difference between quantitative data (based on facts) and qualitative data (based on opinion). Remember: <u>quantit</u>ative relates to <u>quantity</u>, and <u>qualit</u>ative relates to <u>quality</u>.

23

Table 2.1 Comparison of quantitative and qualitative criteria

	Quantitative criteria	Qualitative criteria
Definition	Information that can be expressed as a number and quantified by hard facts	Information that cannot be expressed as a number and describes qualities, opinions and feelings
Criteria type	Numbers and statistics	Words, objects, observations, pictures and symbols
Product specification criteria	Measuring instruments, surveys and testing	Use of the five senses: hearing, taste, smell, sight and touch
Questions that the criteria answer	How much? How many? How often?	How? What? Why?

2.1.3 Reasons for the product criteria included in the design specification (ACCESS FM)

Design specifications contain various product requirements. While ACCESS FM is a useful tool for product analysis, it can also be used to ensure key design requirements are included in the design specification.

Table 2.2 Using ACCESS FM to define design requirements

Categories		Requirement examples	Quantitative/qualitative
A	Aesthetics	+ What should the product look like? + What surface texture should the product have? + What sounds should the product make? + What colour should the product be? + What smell/taste should the product have?	Qualitative
C	Cost	+ How much does it cost to manufacture the product? + Will the product be affordable? + How much will a customer pay for the product? + Will the product be profitable?	Quantitative
C	Customer	+ Who is the product designed for? + What do the customers need from the product? + Who will buy the product? + How will the product influence customers' lives?	Qualitative
E	Environment	+ What are the environmental impacts of the product? + Can the product be manufactured sustainably? + How much recycled material can be used to make the product? + What packaging will be used on the product? + Can the product be recycled, repaired or reused at the end of its life?	Qualitative
S	Size	+ How much should the product weigh? + What size should the product be, for example what are its length, width and height in mm? + Is the product in tolerance? + Should anthropometric data be used?	Quantitative
S	Safety	+ How safe will the product be during use? + Are any harmful materials being used to make the product? + What regulations and standards must the product meet?	Qualitative
F	Function	+ Where will the product be used? + Will the product perform its intended task? + What should be tested to check the product's performance? + What maintenance will be needed during use?	Qualitative
M	Materials	+ What materials will the product be made from? + What materials are available to make the product? + What manufacturing techniques could be used to make the product?	Qualitative

Figure 2.2 What are the key design requirements for a shopping trolley?

2.2 How manufacturing considerations affect design

REVISED ◯

Before a product is made, a designer needs to consider criteria in the design specification that affect manufacture, for example:
+ What materials will be used, and in what form are these materials available?
+ How many parts will be made?
+ What equipment and processes are needed to make the product?

A designer must work closely with the manufacturing engineer, as these requirements affect the design of the product.

2.2.1 Scale of manufacture

Decisions on scale of manufacture are influenced by the quantities required of identical products:
+ One-off production involves making one product at a time. This could be for a specific customer, for example a super tanker for a shipping company or a sculpture to go in a town centre.

25

+ Batch production involves making identical products in groups (batches). Many high-street products are sold in batches, such as furniture and shoes. The appearance and sizes can be changed and adapted between different batches.
+ Mass production involves making large quantities of identical products repeatedly. For example, screws, nuts and bolts are mass produced on a production line.

> **Exam tip**
>
> You need to know typical products made at different scales of manufacture.

> **Typical mistake**
>
> Remember that scale of manufacture refers to the *quantity* of identical products to be made and not to be confused with the size of the products.

One-off production

In one-off production, a large variety of different products can be made using a range of machines and processes. Products are manufactured one at a time and may be large projects, for example tanker ships or unique custom motorcycles. This means that machines and equipment must be able to switch between making new products with different designs, so they are typically operated and controlled by skilled human workers. Capital costs can be low if existing machinery is used, but labour costs per product are high.

> **Capital costs** Fixed, one-time expenses used in the production of goods to purchase buildings, machinery and equipment

Figure 2.3 Examples of one-off production: a lifeboat and a custom racing car

Batch production

In batch production, products or components are manufactured in batches of a set quantity, using a mix of computer- and manually controlled machines. To speed up production, moulds, templates and devices such as fixtures and jigs are often used. The labour costs per product are reduced, as less time is needed in the manufacturing process.

Electronic products such as fridges contain components that need to be made in batches, for example the circuit boards and fridge bodies are made from pressed steel. For different models of fridge, the tools will need to be changed so that the parts can be made in batches.

Check your understanding and progress at **www.hoddereducation.co.uk/myrevisionnotes**

Figure 2.4 Circuit boards tend to be produced in batches

Mass production

+ Machines and equipment are used to make the same product, in large volumes, many times repeatedly. Computer-controlled machinery is used to speed up and automate the process.
+ It is important that the design of the product has limited variation, to allow the computer-controlled machines to repeat the same tasks over time.
+ The labour costs per product are low but the initial costs of setting up the computer-controlled machinery are high. However, when this is balanced against the thousands of products made, the capital cost per product can be low.

> **Automate** To make something operate by automatic CNC machinery or equipment
>
> **Limited variation** Where there is little change in the design of a product

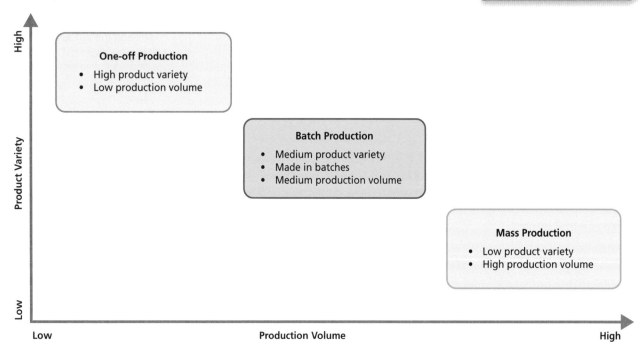

Figure 2.5 Scale of manufacture

2.2.2 Material availability and form

Materials are chosen based on the properties and characteristics required by the product. Most materials are available as a stock form; for example metal is available in bar, rod, plate, sheet and tube. It will also be sold by length, width, thickness and diameter.

Stock form The shape in which materials are available

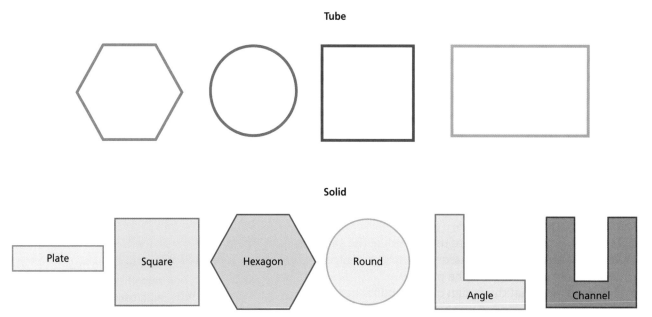

Figure 2.6 Common stock forms of metal

Supply and demand can affect the availability of materials, as this can fluctuate over time. Designers and manufacturers may need to consider using a different, more available material, which could mean a dramatic change in product shape, material properties or manufacturing processes. Businesses also need to consider the types of materials that need to be sourced, for example:
+ raw materials
+ components
+ sub-assemblies
+ tooling
+ packaging
+ consumables (for example oil, lubricant and paint)

Supply and demand Relationship between the quantity of products a business has available to sell and the amount consumers want to buy

Designers often adapt their designs so that they can use stock forms of material. This is cheaper than having material specially made and reduces the amount of waste material cut away, in turn saving money and reducing environmental impact.

2.2.3 Types of manufacturing process

Manufacturing processes change the proportions, shape and size of materials or products in a useful way. There are six types of process (see Figure 2.7), and each will use many different methods and tools.

Typical mistake

Many students suggest an incorrect manufacturing process for a product or component, so make sure you are clear about the different types.

1. Wasting
2. Shaping
3. Forming
4. Joining
5. Finishing
6. Assembly

Figure 2.7 Types of manufacturing process

Check your understanding and progress at **www.hoddereducation.co.uk/myrevisionnotes**

Wasting

Most manufacturing operations begin with material that is larger than the actual part being produced. Machines or tools are then used to remove the unwanted material in order to form the part. This process is known as wasting.

Wasting typically involves:

+ chipping or cutting away excess material using a sharp edge or tool
+ melting material where it needs to be separated.

After wasting, there is less material left on the product. The removed material is then scrap or waste.

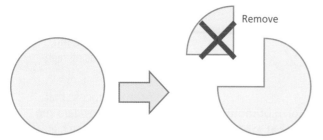

Figure 2.8 Wasting is a process that removes material

To manufacture more complicated shapes, several process steps may be needed. Each feature will need wasting separately and in sequence. This adds time and cost to the manufacturing process. Therefore, when designing products for manufacture, designers will try to use standard forms of material to reduce the wasting processes.

Table 2.3 gives examples of different wasting processes.

Table 2.3 Examples of wasting processes

Wasting process	Definition
Boring	Process of enlarging an existing hole with a cutting tool on a **lathe**
Drilling	Making a hole with a drill
Filing	Rubbing off a small particle with a file when smoothing or shaping
Laser cutting	Cutting with a laser
Milling	Cutting or shaping metal using a rotating tool
Routing	Cutting a groove, or any pattern not extending to the edges, in a wooden or metal surface
Sawing	Cutting using a saw
Shearing	Cutting off with **shears**
Threading	Cutting a screw **thread** in or on a hole, screw, bolt etc.
Turning	Using a **lathe**

Figure 2.9 Laser cutting

Shaping

Shaping processes involve a change in state of a material. This means the material either changes from a liquid to a solid or vice versa. For example, a material might be heated until it melts, then poured into a mould. As the material cools, it solidifies into a new shape.

One benefit of shaping is that a complicated 3D shape can be made in a single process with very little waste.

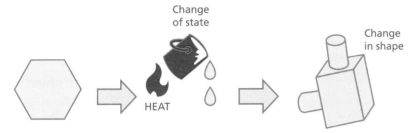

Figure 2.10 Shaping

Most shaping processes, except 3D printing, require some form of mould:
+ Processes such as the sand casting of metals use moulds that are destroyed (to remove the moulded product).
+ Other processes, such as injection moulding of polymers and die casting, use metal moulds. These moulds can be expensive to manufacture and so are typically reused many times.

Table 2.4 gives examples of different shaping processes.

Table 2.4 Examples of shaping processes

Material	Shaping process	Description
Ceramics and metals	Powder metallurgy	Materials or components are made from powders. There are four steps: 1 Powder is made from either metals or ceramics. 2 Powders such as iron, copper and graphite are mixed together. 3 The powder is pressed into shape. 4 **Firing** or **sintering** produces a solid 3D object.
Metals	Die casting	Molten metal is poured or forced into steel moulds.
	Investment casting	Accurate castings are made using a mould formed around a pattern of wax or similar material, which is then removed by melting.
	Sand casting	Molten metal is poured or forced into sand moulds.
Polymers	3D printing	A physical 3D object is made from a 3D digital model, typically by laying down many thin layers of a material in succession.
	Injection moulding	Polymers are shaped by injecting heated material into a mould. There are four steps: 1 **Thermosetting polymers** or **thermoplastic polymers** are melted. 2 The liquid polymer is injected at pressure into a mould cavity. 3 Water cools the mould. 4 The polymer solidifies to produce the final object.

Shaping Process that involves a change in state of a material

Ceramics Materials that are typically an oxide, a nitride or a carbide of a metal

Metals Materials typically made by processing an ore that has been mined or quarried

Firing Process that turns ceramic powders into solid objects using high temperatures

Sintering Process that turns metal powders into solid objects using high temperatures

Thermosetting polymers Polymers that can be shaped and formed when heated, but this process can only occur once; they cannot be reheated or reformed

Thermoplastic polymers Polymers that become pliable when heated, such that they can be shaped and formed, and harden when cooled; this process can be repeated over and over again

Figure 2.11 Shaping by sand casting

Forming

Forming involves changing the shape of a material without changing its state. It includes processes such as press forming and rolling, which bend or deform material (see Figure 2.13).

Some forming processes also heat the material, such as extrusion, to make it easier to change its shape, but the material never reaches its melting point.

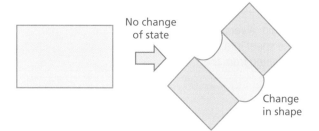

No change of state

Change in shape

Figure 2.12 Forming

> **Forming** Process that changes the shape of a material without changing its state
>
> **Extrusion** Forming process in which a metal or polymer is forced through a die
>
> **Melting point** Temperature at which a substance changes from a solid to a liquid

Figure 2.13 Press-forming metal tube

31

The forming process only produces simple shapes, so designers are limited with their design choices. After a material has been formed, it will often need a further wasting process to remove any excess material from the formed shape.

Table 2.5 gives examples of forming processes.

Table 2.5 Examples of forming processes

Material	Forming process	Description
Composites	Moulding	A binding liquid (e.g. epoxy resin) is forced into the reinforcing material (e.g. carbon fibre fabric).
Metals	Forging	Metal is shaped using localised **compressive forces**. The blows are typically delivered with a power hammer.
	Press forming	A shape is formed from a metal sheet or tube, using a press tool and stamping **die**. The metal is manipulated to the shape of the die to produce precise and accurate components.
	Rolling	Metal stock is passed through one or more pairs of rollers to reduce the thickness.
Metals/polymers	Extrusion	Objects of a fixed cross-sectional profile are created by pushing material through a die of the desired cross-section.
Polymers	Strip heating	A polymer is heated along a line so that it becomes soft and flexible. It can then be folded to almost any angle.
	Vacuum forming	A polymer sheet is heated until it is soft and then draped over a mould. A vacuum is then applied, sucking the sheet into the mould. The sheet is then ejected from the mould.

Joining

Joining processes attach separate pieces of material together to make a product. They allow products to be made from several components made from more than one type of material.

Joining processes can be either permanent or temporary:
+ Permanent joints cannot be dismantled without breaking the component parts.
+ Temporary joints can be dismantled without breaking the component parts and are useful where disassembly is required.

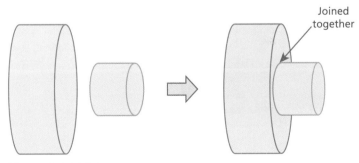

Joined together

Figure 2.14 Joining

Composites Materials consisting of two (or more) different materials bonded together

Compressive forces Pressing or squeezing forces

Die Tool that imparts a desired shape on a metal or polymer; typically a perforated block through which a metal or polymer is forced by extrusion

Joining Process that attaches separate pieces of material together to make a product

Table 2.6 shows the differences between permanent and temporary joints.

Table 2.6 Differences between permanent and temporary joints

	Permanent joint	Temporary joint
Dismantling	Joint cannot be dismantled without damaging the assembled parts.	Joint allows easy dismantling of the assembled parts without breaking them.
Strength	Strength of a permanent joint is high. Joint strength is the same as the parts.	Strength of a temporary joint is comparatively lower than the strength of the parts.
Repair and replacement	Repair and replacement are difficult and costly.	Repair and replacement are easy.
Leakage	Permanent joints are usually leak-proof.	Temporary joints are not necessarily leak-proof.
Permanent/ temporary	Permanent joints are suitable where products do not require separating.	Temporary joints are suitable where products need to be separated frequently.
Examples	Soldering Brazing Welding Press forming Adhesive joining	Threaded fasteners Press fit Rivets Cotter pins

Designers must decide what properties they are looking for in a joining method. This is because the joint may have different properties to the rest of the material.

Figure 2.15 Welding is an example of a permanent joining process

Finishing

Finishing processes change the surface of a material to make it useful; this may involve a colour or texture change.

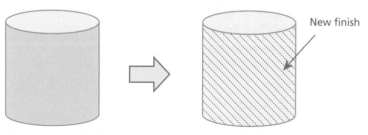

New finish

Figure 2.16 Finishing

Soldering Joining different types of metal by melting solder

Brazing Joining metal parts by melting and flowing a filler metal into the joint, using heat provided by either a flame or an oven

Welding Joining metal parts where the part edges are melted by means of heat

Press forming Forming operation used to change the shape of a sheet material using pressure from a press

Threaded fasteners Products such as screws, nuts and bolts

Press fit Mechanical, solderless method of tube joining

Rivets Mechanical fasteners with a head on one end and a cylindrical stem on the other

Cotter pins Pin-shaped pieces of metal that pass through a hole to fasten two parts of a mechanism together

Finishing Process that changes the surface of a material in a useful way

33

Many finishing processes are additive:

+ Electroplating can add a layer of expensive metal to a cheaper metal. The benefit is that the product does not need to be made wholly of the expensive metal, therefore saving money and improving functionality. Electroplating electronic connectors in gold enhances electrical conductivity.
+ Adding layers of paint to a product (see Figure 2.17) may protect it from corrosion, while changing the colour improves its decorative appeal.

Additive Where material is added to create a product, for example 3D printing

Electroplating Process that produces a metal coating on a part by placing it in a chemical bath and passing an electrical current through it

Assembly Process of fitting component parts together to make a whole product

Tolerance Amount of variation allowed in a dimension if the product is to work as intended

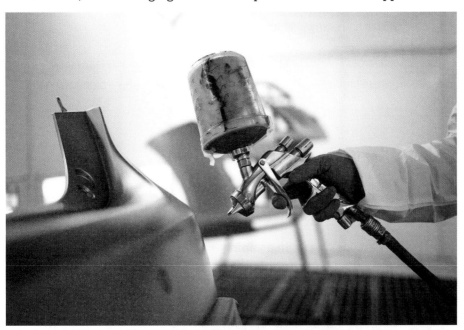

Figure 2.17 Paint spraying is an example of a finishing process

Table 2.7 shows examples of finishing processes.

Table 2.7 Examples of finishing processes

Material	Finishing process	Description
Most materials	Painting	Applying paint to an object using either a spray or a brush
	Lacquering	Applying liquid made of shellac dissolved in alcohol, which dries to form a hard protective coating
Metals	Electroplating	Coating a metal object by electrolytic deposition with chromium, silver or another metal
	Ceramic and powder coatings	Covering an object with a plastic (or ceramic mix) powder, which is then heated to flow and fuse into a coating
Woods	Dyes	Using a dye to penetrate wood and colour it from within

Assembly

Assembly is the process of fitting component parts together to make a whole product. This can involve some joining processes, such as press fit, fasteners and soldering, as well as pushing, sliding and aligning parts together. Depending on the scale of production, assembly lines may be fully automated using robots and CNC machinery or assembly may be carried out by humans.

Designers must consider how parts will be assembled to ensure that features such as holes and edges align. They will keep control of the product design by stating a tolerance for every dimension; this is the acceptable amount of variation allowed on a dimension to ensure a product can fit together and function properly.

For example, the length of a metal bar might need to be cut to 30 mm. If the tolerance is ±0.5 mm, then anything outside the range of 29.5–30.5 mm would be unacceptable. This is important, as if tolerances are not met, parts of products may not fit together properly. This would then involve more time and money to remake the parts.

Check your understanding and progress at **www.hoddereducation.co.uk/myrevisionnotes**

Figure 2.18 Assembly-line production

2.2.4 Production costs

Production costs are all the expenses a company must pay in order to manufacture a product. These costs will affect manufacturing design considerations.

Manufacturers focus on the costs of:
+ labour to make the product
+ equipment and processes
+ materials
+ factory overheads, such as rent, heating, insurance and power.

> **Overheads** Costs of running a business, including rent, insurance and utilities

Check your understanding

6 Name **one** scale of manufacture.
7 Give **two** forms of supply for tube or solid metal.
8 Identify the type of process described by the following statements:
   a Bending a sheet of metal
   b Melting a material and pouring it into a mould.
9 What is meant by the term legislation?

Now test yourself

TESTED ◯

Many different finishes can be used on a variety of materials.

1 Identify five objects around your house that have some form of finish. Why has this type of finish been used?
2 One type of finish is electroplating. Research this process and then draw and label a diagram of the key components of how the process works.

Exam tip

Make sure you know your manufacturing processes. The ability to look at a product and recognise the manufacturing process will save lots of time and lead to more accurate answers in the exam.

2.3 Influences on engineering product design

There are many influences on engineering product design. Not only do designers have to think about how a product will be made, but they must also consider a range of issues such as:

+ What market is there for a new product?
+ What new technologies are available?
+ What standards and legislation must be followed?
+ How can a product be made sustainable and action the circular economy?

2.3.1 Market pull and technology push

Market pull describes when a need for a product arises from customer demand. The need is identified through market research and products are developed to satisfy the need.

Digital cameras are an example of market pull. They have developed over the years to meet the changing demands of professional photographers (the customer):

+ Figure 2.19 shows a camera from the 1980s that uses a roll of film that will take 36 photographs.
+ The market needed cameras that could store large amounts of photographs, and this came about with the development of digital cameras (Figure 2.20).
+ Digital cameras can store thousands of photographs on a memory card, with a battery life that allows the camera to be used for a long time without recharging.

> **Market pull** When a need for a product arises from customer demand, which 'pulls' development of the product

Figure 2.19 A 35 mm camera and roll of film

Figure 2.20 A digital camera and storage card

Check your understanding and progress at **www.hoddereducation.co.uk/myrevisionnotes**

Technology push is when new technology is created and new products are developed. An example is touchscreen technology, which has changed the way we use our electronic devices.

The first touchscreen computers were designed in the 1980s to replace the use of dials and buttons. Today, we can operate smartphones, smartwatches and a range of devices in the home with touchscreen technology.

Technology push When new technology is created as a result of research and development (R&D), resulting in new products that are 'pushed' into the market, with or without demand

Standards Guidelines on how to meet legislation that set out agreed ways of doing something, such as making a product, managing a process or supplying materials

Figure 2.21 Smartwatch using touchscreen technology

2.3.2 British and international standards

Standards are guidelines on how to meet legislation. They set out agreed ways of doing something and are a statement of good practice, designed to make things better and safer.

British Standards (BS) are produced by the British Standards Institution (BSI), which is the national standards body of the United Kingdom (UK).

BSI Kitemark

The BSI Kitemark is a quality mark owned and managed by BSI. It is used in the UK but recognised internationally. Products that satisfy the Kitemark scheme can display the BSI Kitemark logo.

European Conformity (CE) marking

International standards include CE marking. This indicates that a product has been assessed by the manufacturer and is deemed to meet European Union (EU) safety, health and environmental protection requirements. It is required for products manufactured anywhere in the world that are then marketed in the EU.

United Kingdom Conformity Assessment (UKCA)

The United Kingdom Conformity Assessment (UKCA) marking came into effect on 1 January 2021. It is the new UK product marking for products being placed on the market in Great Britain (England, Wales and Scotland) and is not recognised in the EU. Products that require CE marking still need CE marking to be sold in the EU.

2.3.3 Legislation

Legislation means laws proposed by the government and made official by Acts of Parliament. Some laws are designed to protect the user or consumer, by ensuring products comply with relevant safety standards. A designer could face legal action if a product was found to be unsafe or have caused harm. If a company or an individual is shown to have broken the law, they can be prosecuted.

One example of legislation in the UK is the Health and Safety at Work etc. Act 1974 (HASAWA), which covers a wide range of duties for both employers and employees in the workplace. HASAWA specifically states that an employer must provide risk assessments to keep all people at work safe from workplace hazards. The employer must also provide health and safety training and ensure the control of dangerous substances and emissions.

> **Legislation** Laws proposed by the government and made official by Acts of Parliament
>
> **Prosecuted** Officially accused in a court of breaking the law

2.3.4 Planned obsolescence

Planned obsolescence is a policy of producing consumer goods that rapidly become obsolete (out of date and therefore unusable), either due to changes in design or unavailability of spare parts. This policy is adopted because large corporations want customers to buy more products, so they can make more profit. The effect of this policy is the creation of waste and emissions.

> **Planned obsolescence** Policy of producing consumer goods that rapidly become out of date and unusable, either due to changes in design or unavailability of spare parts

Examples of planned obsolescence include:
+ limiting the life of a lightbulb
+ launching a new model of a car every year with minor changes
+ using irreplaceable batteries in tech products
+ being unable to replace an ink cartridge in a printer.

Figure 2.22 Printers can be made obsolete by stopping manufacture of compatible printer inks

Check your understanding and progress at **www.hoddereducation.co.uk/myrevisionnotes**

2.3.5 Sustainable design (6 Rs)

Designers need to consider the effect that products have on the environment:
+ Many of the products we use daily contain materials that are in scarce supply and non-renewable.
+ A lot of energy is used during raw material extraction and material processing, as well as during the manufacture, transport, use and disposal of products.
+ The energy used throughout the product life cycle releases carbon dioxide, which contributes to climate change.

The 6 Rs of sustainable design (see Table 2.8) help designers to reduce the impact of a new product.

Table 2.8 The 6 Rs of sustainable design

6 Rs	Description
Reduce	Cut down on the amount of material used as much as possible.
Reuse	Use a product to make something else with all or parts of it.
Recycle	Reprocess a material or product and make something else.
Rethink	Design in a way that considers people and the environment.
Refuse	Do not use a material if it is bad for the environment.
Repair	When a product breaks down or stops working properly, fix it.

Carbon dioxide Colourless and non-flammable gas that exists in the Earth's atmosphere; rising carbon dioxide levels are contributing to global warming

Sustainable Using something in a way that ensures it does not run out and affect the needs of future generations

Exam tip

Some questions might ask for a definition of one of the 6 Rs of sustainable design, so make sure you know and understand them.

2.3.6 Design for the circular economy

In our current economy, we take materials from the Earth, make products from them, and eventually throw them away as waste. This process is linear – a series of steps progressing from one to the other, from beginning to end.

The circular economy, by contrast, is a system designed to reduce the impact of the use of materials on the environment by:
+ reducing waste and pollution
+ reusing and recycling materials.

It challenges designers to consider the impact a product will have throughout its life cycle, since we live in a world with finite (limited) materials.

Circular economy System designed to reduce the impact of the use of materials on the environment by reducing waste and pollution and reusing and recycling materials

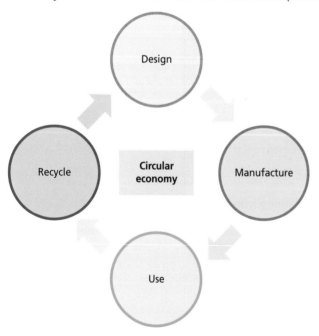

Figure 2.23 The circular economy

Check your understanding

10 What is meant by 'technology push'?

11 What are 'standards'?

12 Describe what is meant by 'planned obsolescence'.

13 Draw a diagram to show the four parts of the circular economy.

Exam tip

Do not concentrate solely on recycling when asked how products can be made more sustainable. Try to include a range of issues, such as choice of materials, efficiency of distribution, and potential for making designs from individual components to facilitate easier repair or upgrades.

Now test yourself

TESTED ◯

Use the 6 Rs of sustainable design to identify what impact a children's party pack has on the environment. Think about the materials used, the packaging, and how waste is disposed of.

Figure 2.24 Children's party pack: what impact does it have on the environment?

Exam-style questions

1 Describe what is meant by a product need. [1]

2 Which **one** of the following options gives examples of quantitative criteria? [1]

  a Pictures and symbols ☐

  b Numbers and statistics ☐

  c Hearing and taste ☐

  d Words and opinions ☐

3 State **one** example of the requirements for cost in ACCESS FM. [1]

4 Identify **one** example of a finishing process for metals. [1]

5 State **one** production cost that a company must pay. [1]

6 Explain **two** influences that a designer must consider when designing a new product. [4]

7 Describe how legislation protects the users of products from harm. [3]

8 Describe, with **one** example, planned obsolescence. [4]

9 State **one** of the 6 Rs of sustainable design and describe its meaning. [2]

10 Describe what is meant by the circular economy. [3]

Check your understanding and progress at **www.hoddereducation.co.uk/myrevisionnotes**

Exam checklist

In this topic, you learned about the following:

+ Criteria included in an engineering design specification:
  + Needs and wants
  + Quantitative and qualitative criteria
  + Reasons for the product criteria included in the design specification (ACCESS FM: Aesthetics, Cost, Customer, Environment, Size, Safety, Function, Materials)
+ How manufacturing considerations affect design:
  + Scale of manufacture: one-off, batch, mass
  + Typical products manufactured at different scales of manufacture
  + Material availability and form
  + Types of manufacturing process (wasting, shaping, forming, joining, finishing, assembly)
  + Production costs: labour, capital
+ Influences on engineering product design:
  + Market pull and technology push
  + Quality standards, such as British and international standards and United Kingdom Conformity Assessment (UKCA)
  + Legislation related to health and safety regulation and risk assessment
  + Planned obsolescence
  + The 6 Rs of sustainable design: Reduce, Reuse, Recycle, Rethink, Refuse, Repair
  + Design for the circular economy.

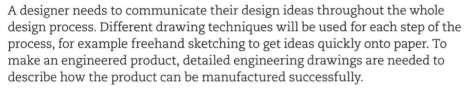

Topic area 3: Communicating design outcomes

3.1 Types of drawing used in engineering REVISED ●

A designer needs to communicate their design ideas throughout the whole design process. Different drawing techniques will be used for each step of the process, for example freehand sketching to get ideas quickly onto paper. To make an engineered product, detailed engineering drawings are needed to describe how the product can be manufactured successfully.

The design drawing process is like a relay race. Freehand sketching is the early part of the race that sets up the foundations for the success of a product. The drawings will adapt as the race develops to show more information and detail, until it ends with formal engineering drawings.

3.1.1 Freehand sketching

When designers are thinking about new products, they will use freehand sketching to quickly show others their initial ideas. All that is needed is a drawing tool, typically a pencil or a pen, and something to draw on.

A sketch is a rough drawing that can communicate concepts without having to be formally drawn or completely accurate. Once designers have got their ideas on paper, it makes it easier to spot potential problems before any prototypes are made.

> **Freehand sketching**
> Drawing without the use of measuring instruments

Figure 3.1 Freehand sketching

3.1.2 Isometric drawing

Isometric drawing, also known as isometric projection, gives depth to a sketch by showing it in three dimensions (3D):

+ It presents a front-edge view of the nearest corner followed by lines of sight at 30 degrees to the horizontal (see Figures 3.2 and 3.3).
+ It uses the same scale at each side of the object.
+ Vertical lines remain vertical.

Both designers and engineers use and understand this method.

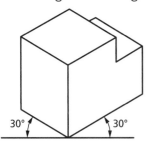

Figure 3.2 Isometric drawing angles

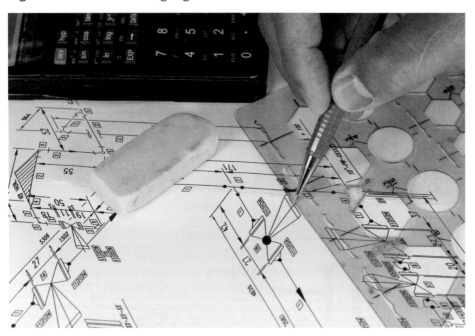

Figure 3.3 Isometric drawing

3.1.3 Oblique drawing

Also known as oblique projection, oblique drawing is a formal technique that ensures information is understood by other people. It is used to develop a two-dimensional (2D) sketch and show it in 3D:

+ An initial 2D sketch is drawn.
+ The lines of sight are drawn at 45 degrees to the horizontal.
+ Vertical lines remain vertical.
+ The depth of the drawing can vary from full depth to a scaled view.

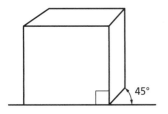
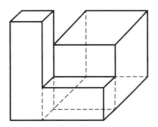

Figure 3.4 Oblique drawings

3.1.4 Orthographic drawing

An orthographic drawing is a formal working drawing that shows an object from every angle, to help manufacturers plan production. It represents a 3D object by using several 2D views of it. The elevations (sometimes called projections) are front, side and plan (top). They are drawn to scale and must show dimensions (see Figure 3.5).

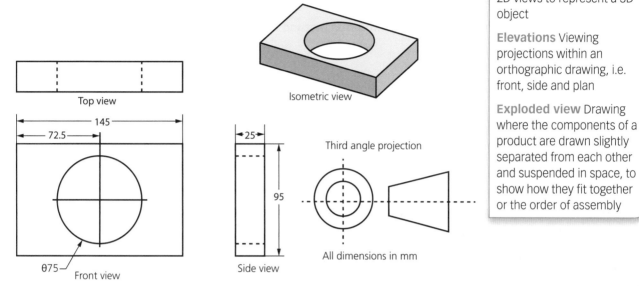

Figure 3.5 Orthographic drawing including an isometric view

In Figure 3.5, the symbol (circles and a cone) denotes that this is a third angle orthographic drawing. The drawing also includes accurate dimensions (in millimetres), which allows the product to be manufactured.

3.1.5 Exploded views

An exploded view is a drawing where the components of a product are drawn slightly separated from each other and suspended in space, to show how they fit together or the order of assembly. They are drawn in isometric and the parts are arranged so that each can be clearly identified while maintaining positions in relation to each other.

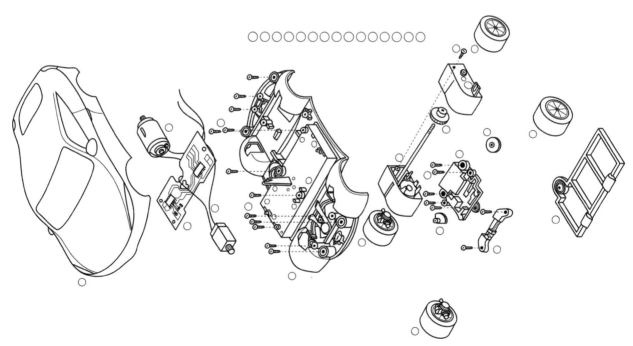

Figure 3.6 Exploded view

Orthographic drawing
Formal working drawing that shows an object from every angle, to help manufacturers plan production; it uses several 2D views to represent a 3D object

Elevations Viewing projections within an orthographic drawing, i.e. front, side and plan

Exploded view Drawing where the components of a product are drawn slightly separated from each other and suspended in space, to show how they fit together or the order of assembly

Check your understanding and progress at **www.hoddereducation.co.uk/myrevisionnotes**

3.1.6 Assembly drawings

An assembly drawing gives the big picture of a completed project. It shows how separate components are joined together to make a whole product.

The most common assembly drawing is called a general assembly, which:
+ is drawn in isometric
+ shows the general arrangement of the parts
+ provides a parts list and parts numbers.

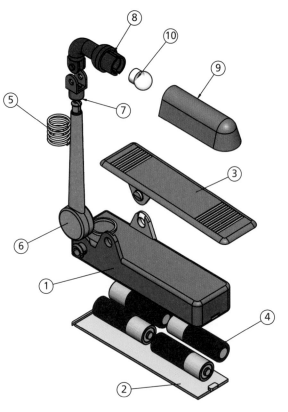

ITEM NO.	PART NAME	QTY.
1	Base	1
2	Cover	1
3	Clamp	1
4	Battery	4
5	Spring	1
6	Arm	1
7	Swivel	1
8	Elbow	1
9	Shade	1
10	Bulb	1

Figure 3.7 Assembly drawing

3.1.7 Block diagrams

Block diagrams are drawings of systems or products where separate parts are represented by blocks and connected by lines showing their relationship to one another.

Block diagrams:
+ give a better understanding of how a system functions and the interconnections within it
+ are generally used in electronics and computing, for the design and analysis of a system.

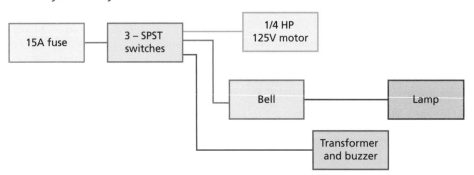

Figure 3.8 Block diagram

3.1.8 Flowcharts

Flowcharts are block diagrams that show how various processes are linked together to achieve a specific outcome. They:
+ show a step-by-step approach to solving a task
+ use various kinds of boxes – typically rectangular boxes for processes and diamond boxes for decisions
+ show the order of a workflow or process by connecting the boxes with arrows.

Flowcharts Block diagrams that show how various processes are linked together to achieve a specific outcome

Circuit diagrams Conventional graphical representations of electrical circuits, where separate electrical components are connected to one another

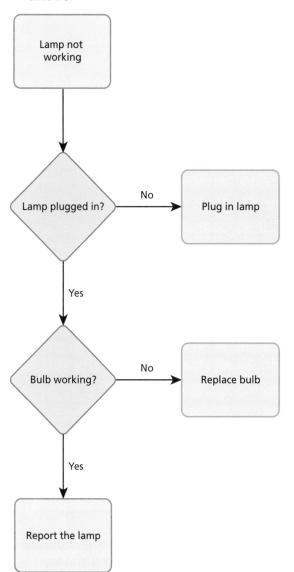

Figure 3.9 Example of a flowchart

3.1.9 Circuit diagrams

Circuit diagrams are conventional graphical representations of electrical circuits, where separate electrical components are connected to one another. They are used for the design, construction and maintenance of electrical and electronic equipment.

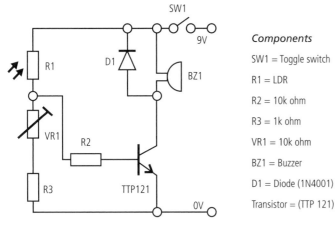

Components

SW1 = Toggle switch

R1 = LDR

R2 = 10k ohm

R3 = 1k ohm

VR1 = 10k ohm

BZ1 = Buzzer

D1 = Diode (1N4001)

Transistor = (TTP 121)

Figure 3.10 Example of a circuit diagram

3.1.10 Wiring diagrams

Wiring diagrams are simplified pictorial representations of circuits that show how each component should be connected.

> **Wiring diagrams** Simplified pictorial representations of circuits that show how each component should be connected

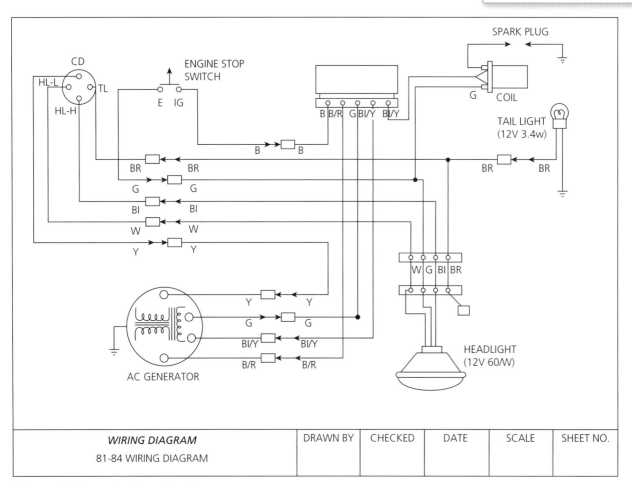

Figure 3.11 Example of a wiring diagram

Exam tip

Examiners may ask you to identify a type of drawing by looking at an image. Make sure you know what each type of drawing looks like and any key features.

3.1.11 Advantages and disadvantages of each drawing technique

Table 3.1 details the relative advantages and disadvantages of each drawing technique.

Table 3.1 Advantages and disadvantages of types of drawing technique

Drawing technique	Advantages	Disadvantages
Freehand sketching	Enables an idea to be visualised quickly Uses a universal language that all designers understand	Drawing by hand is slower than other methods Difficult to correct mistakes
Isometric drawing	Provides an overall 3D view of a product Used by designers and engineers to show a realistic image	Requires a high level of skill to produce readable drawings Cannot see the rear of the object
Oblique drawing	Easy to draw Clear front view of the object	Difficult to dimension Looks less realistic than isometric drawing Cannot see the rear of the object
Orthographic drawing	Universal standard used by all engineers All information clearly laid out	2D limits the realism of the drawing Untrained readers may struggle to visualise the object in 3D and interpret features and symbols
Exploded view	Helps understanding of how parts are arranged	Cannot see the rear of the object
Assembly drawing	Helps understanding of how parts are assembled	Cannot see the rear of the object
Block diagram	Shows the function of a system Easy to construct and read	No information about the physical construction of a system
Flowchart	Shows the logic of a system Easy to understand and follow	Modification can be time consuming Showing many processes is difficult
Circuit/wiring diagram	Shows clearly how circuits are assembled	Readers need to understand the symbols used

Check your understanding

1. What is a block diagram?
2. What are exploded views used for?
3. What are flowcharts used for?
4. Name **one** advantage of freehand sketching.
5. Name **one** disadvantage of oblique drawing.

Now test yourself TESTED ◯

1 Figure 3.12 is a simple line drawing showing the front and side views of a handheld radio, along with its basic overall size.

Produce isometric and oblique views of the handheld radio. You can make up your own dimensions to construct the other features, such as the display, side talk button and keypad.

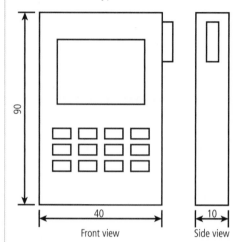

Front view   Side view

Measurements in mm

Figure 3.12 Simple line drawing of a handheld radio

2 An influencer wants a different look when creating media content. Develop some concept sketches for new computer equipment as follows:
   + a webcam
   + a computer mouse
   + a microphone.

3.2 Working drawings

REVISED ◯

3.2.1 2D engineering drawings using third angle orthographic projection

Orthographic projections combine front, side and plan views. The views are always aligned and dimensioned. They provide all the information needed to manufacture the object and are therefore called working drawings.

3.2.2 Standard conventions

Orthographic drawings are drawn to national and international standards. This is so that anyone picking up the drawings can understand fully what they are trying to communicate.

In the UK, British Standard 8888 is the standard convention for engineering drawing. It ensures that all designers and manufacturers use and understand the same conventions. It defines clearly, by reference to several documents, what types of line should be used, how to represent dimensions and how to indicate certain mechanical features, such as holes, chamfers and screw threads. For example, all drawing abbreviations are found in BS 8888: *Technical product documentation and specification* (*Annex A: BIP 2155 The essential guide to technical product specification – Engineering drawing*).

> **Dimensioned** Marked with numerical values to communicate the sizes of key features (measurements are usually in millimetres)

> **Typical mistake**
>
> Students often confuse standard conventions with working drawing types. Remember that standard conventions give the rules, while working drawing types show the application of the rules.

Exam tip

Remember that a standard convention is an agreed set of rules, and that BS 8888 is used so that all designers and manufacturers understand and follow the same rules.

Drawing layout

Working drawings:
+ have a border that separates the design on the paper
+ often have grid references to allow easy identification of areas of the drawing (for example in Figure 3.13, the centre of the circle is in C4).

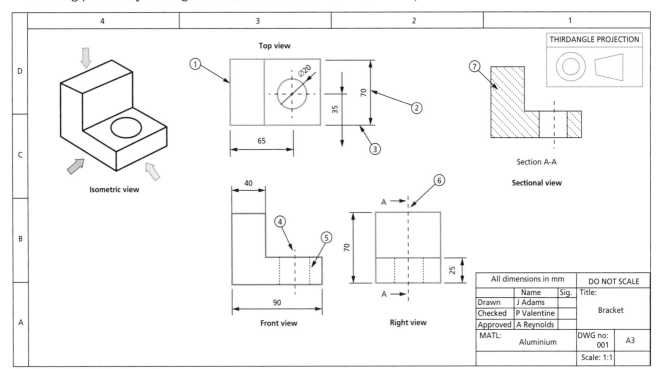

Figure 3.13 Working drawing border with grid reference

Title block

The title block is where all the drawing information is recorded. This ensures the drawing can be identified, interpreted and then archived. It typically appears in the bottom right of the border, as shown in Figure 3.13.

DO NOT SCALE DRG			ALL DIMENSIONS IN MM	
MAT: BRASS			TITLE: HINGE	
HINGE	NAME	DATE	DRG NO: 002	
DRAWN	M Smith	12/10	SCALE:	1:1
CHECKED	G Jakes	13/10	SHEET:	1 OF 1
APPROVED	S Hall	20/10	REV:	2

Figure 3.14 Title block

Metric units of measurement

Units of measurement in the UK are metric, and dimensions in drawings are usually marked in millimetres (mm).

Scale

Scale is used by designers to represent the actual size of a product in their working drawings. 'Do not scale' is often included in drawings, which means that dimensions cannot be measured directly off the drawing itself.

Commonly used scales are:
+ 1:2 – half the actual size
+ 1:1 – actual size
+ 2:1 – twice the actual size.

The scale in Figures 3.13 and 3.14 is given as 1:1.

> **Scale** Amount by which a drawing is enlarged or reduced from the actual size of an object, shown as a ratio

Tolerance

Tolerance is the acceptable amount of variation allowed on a dimension. The designer works out the tolerance based on form, fit and function of a part. It determines how well a part will fit in the final piece and how reliable the final product will be.

If a dimension is within tolerance, it is accurate enough to ensure the product can fit together and function properly. For example, a 120 mm bar with a tolerance of ±0.20 would give an allowance of 119.8 mm to 120.2 mm when the component is manufactured, within which it is still considered accurate.

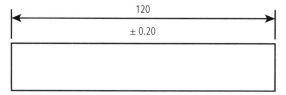

Figure 3.15 Tolerance

Dimensions

Dimensions on working drawings follow a standard convention to ensure manufacturers understand the product requirements created by the designer.

For linear measurements:
+ Dimension and projection lines are thinner than drawing lines.
+ Dimension lines should touch the projection line with a small filled-in arrowhead at each end (see Figure 3.16).
+ The shortest dimensions are placed closest to the drawing. Longer dimensions are positioned further away.
+ Measurements, typically in millimetres (mm), are written in the middle of each dimension line.
+ Measurements are written above the dimension lines, such that they can be read from the bottom of the drawing or from the right only when the drawing is turned around (see Figure 3.16).

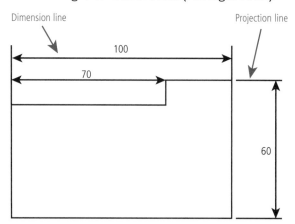

Figure 3.16 Dimension and projection lines

Measurement of a circle or an arc is prefixed with either:
+ R to show the radius
+ Ø to show the diameter.

The arrow head:
+ if external, always points towards the centre
+ if internal, always points away from the centre.

Radius A straight line from the centre of a circle to its circumference (outer edge)

Diameter A straight line from one side of a circle to the other through the centre

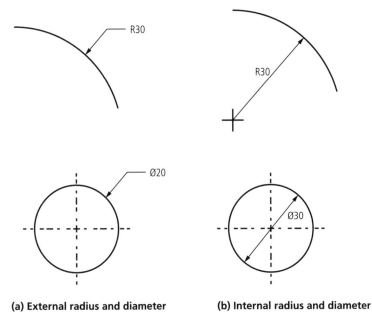

(a) External radius and diameter　　　**(b) Internal radius and diameter**

Figure 3.17 Radius and diameter measurements

In practice, designers would use the radius and diameter projection lines shown in Figure 3.18.

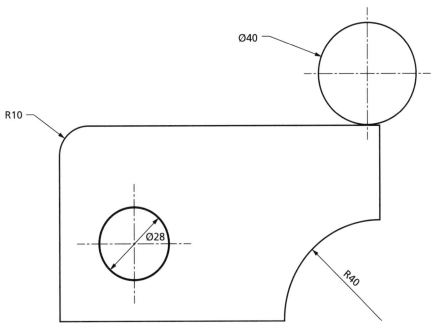

Figure 3.18 Radius and diameter projection lines

Check your understanding and progress at **www.hoddereducation.co.uk/myrevisionnotes**

Surface finish

Designers use surface finish symbols to advise manufacturing engineers on the machined finish a product should have. Figure 3.19 shows the different types of symbol used.

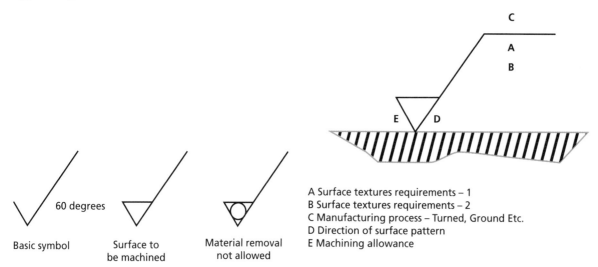

A Surface textures requirements – 1
B Surface textures requirements – 2
C Manufacturing process – Turned, Ground Etc.
D Direction of surface pattern
E Machining allowance

Basic symbol   Surface to be machined   Material removal not allowed

Figure 3.19 Surface finish standards

A designer would use the symbol shown in Figure 3.20 to state a surface finish of 3.2 µm Ra, where µm represents micrometres and Ra is the measurement used for surface finish.

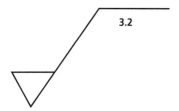

Figure 3.20 Symbol for a surface finish of 3.2 µm Ra

Meaning of line types

Different types of line have different meanings in orthographic drawing, as shown in Table 3.2.

Table 3.2 Types of line in orthographic drawing

Line type	Use
Outlines (thick) ▬▬▬▬▬	To show object outlines and visible edges
Projection lines (thin) ————	To extend lines that assist in 2D drawing
Dimension lines (thin) ←——————→	To show the dimensions of an object
Hidden details (thin) – – – – – – – ·	To show the existence of a hidden edge
Centre lines (thin) ▬ · ▬ · ▬ · ▬ · ▬ · ▬	To show lines that pass through the centre of an object
Break lines (thin) ∿ or ⌁	To show a break in an object

Leader lines connect a graphical representation on a drawing to some text. Different types of leader line terminator may be used, as shown in Table 3.3.

> **Leader line terminator**
> The type of shape used to end a line

Table 3.3 Leader line terminators

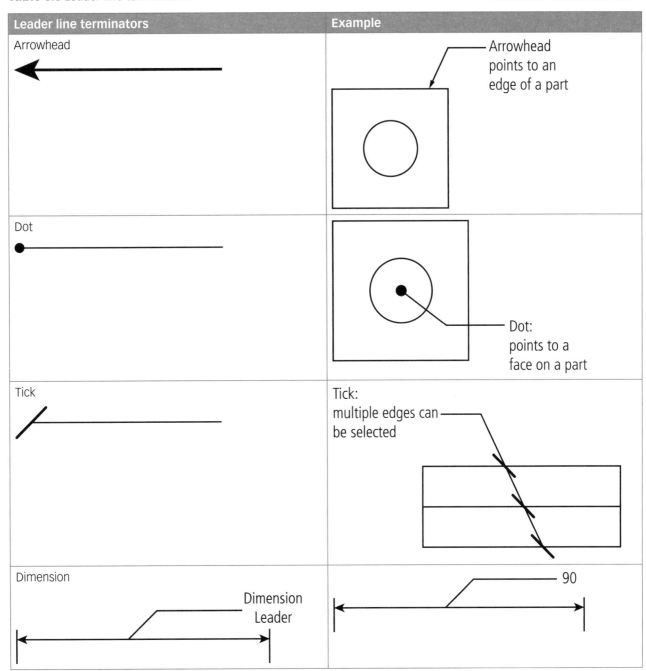

Leader line terminators	Example
Arrowhead	Arrowhead points to an edge of a part
Dot	Dot: points to a face on a part
Tick	Tick: multiple edges can be selected
Dimension	Dimension Leader — 90

3.2.3 Abbreviations

Designers use standard abbreviations to communicate their drawing requirements to manufacturers, as shown in Table 3.4. The full list of abbreviations is in BS 8888: Technical product documentation and specification (*Annex A: BIP 2155 The essential guide to technical product specification – Engineering drawing*).

Table 3.4 Standard abbreviations

Term	Abbreviation	Application	Example
Across flats	AF	Width across flats is the distance between two parallel surfaces on the head of a screw or bolt, or a nut as shown.	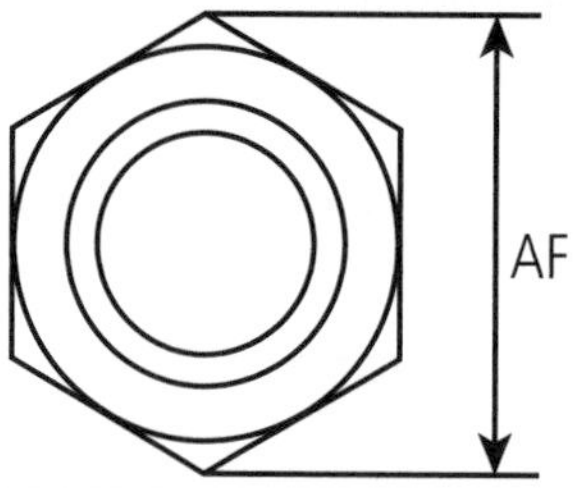 AF
Centre line	CL	A centre line is used to show the midpoint of a feature, such as the centre of a hole.	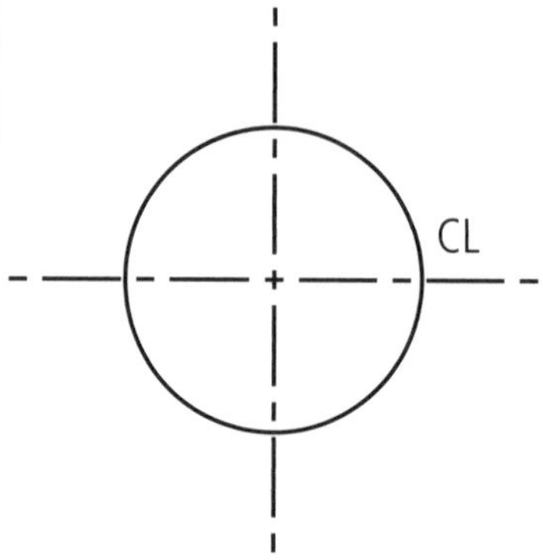 CL
Diameter (in a note)	DIA	This is used in the text notes on a drawing to indicate the diameter of a feature.	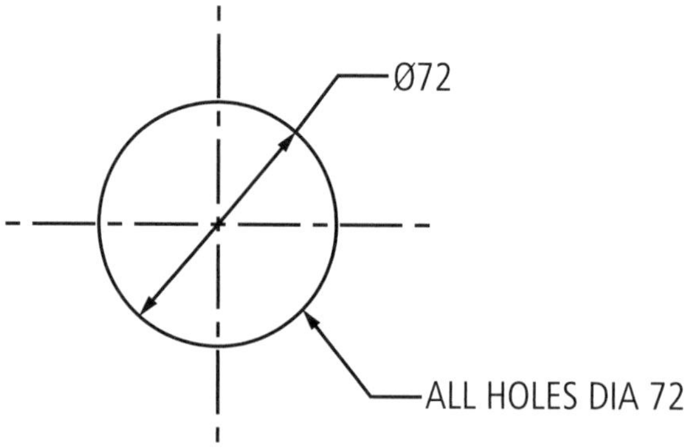 Ø72 ALL HOLES DIA 72
Diameter (preceding a dimension)	Ø	The diameter of a feature, such as a hole, is represented by the Ø symbol.	
Drawing	DRG	This is the shorthand for 'drawing'.	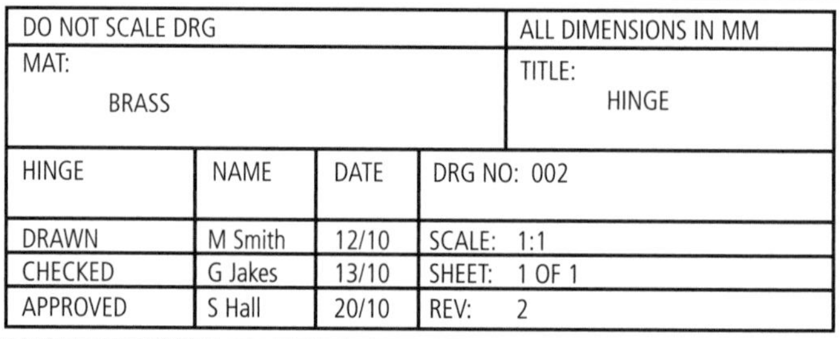 DO NOT SCALE DRG — ALL DIMENSIONS IN MM MAT: BRASS — TITLE: HINGE HINGE / NAME / DATE / DRG NO: 002 DRAWN / M Smith / 12/10 / SCALE: 1:1 CHECKED / G Jakes / 13/10 / SHEET: 1 OF 1 APPROVED / S Hall / 20/10 / REV: 2
Material	MAT	This is the shorthand for 'material'.	
Square (in a note)	SQ	This is used in the text notes on a drawing to indicate a square feature.	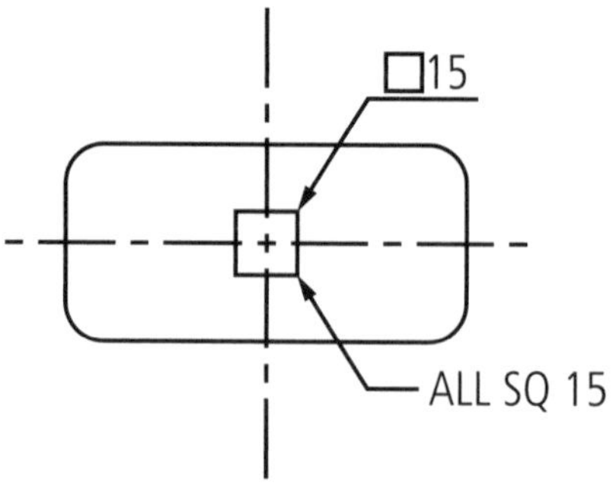 □15 ALL SQ 15
Square (preceding a dimension)	□	This is used to indicate the dimensions of a square feature.	

Exam tip

You need to remember the list of abbreviations shown in Table 3.4. You could make revision cards to aid your memory and note that each abbreviation is a shortening of the word.

3.2.4 Representations of mechanical features

Designers use a wide range of symbols to advise the manufacturer what mechanical features are required for a product.

Threads and chamfers

An internal thread is drawn as shown in Figure 3.21. An external thread and chamfer are drawn as shown in Figure 3.22.

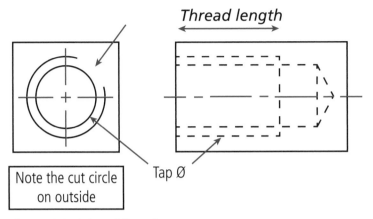

Thread length

Tap Ø

Note the cut circle on outside

Figure 3.21 Internal thread

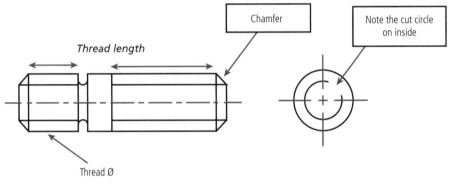

Chamfer

Note the cut circle on inside

Thread length

Thread Ø

Figure 3.22 External thread and chamfer

Holes

Holes are shown as a solid circle or using dashed hidden detail lines. Blind holes and through holes are drawn as shown in Figure 3.23.

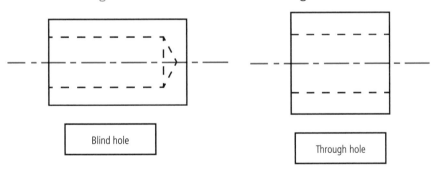

Blind hole

Through hole

Figure 3.23 Blind and through holes

Countersinks

Countersinks (Figure 3.24) are conical holes cut into an object at a given angle so that a bolt or screw can be sunk below the surface.

Thread Spiral grooves of equal measurement on a cylindrical article or pipe; threads can be on the inside (nut) or outside (screw) of a cylinder

Chamfer A transitional edge (or cut-away) between two faces of an object

Blind holes Holes that do not break through to the other side of the workpiece

Through holes Holes that break through to the other side of the workpiece

Countersinks Conical holes cut into an object at a given angle so that a bolt or screw can be sunk below the surface

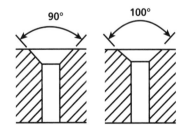

90° 100°

Figure 3.24 Countersinks

Knurls

Knurling is a process of machining a series of straight or criss-cross ridges around the edge of something so that it is easier to grip (such as a control knob).

Knurls are shown on an engineering drawing as a series of straight or criss-cross lines. Straight and diamond knurling are shown in Figure 3.25.

> **Knurling** Process of machining a series of straight or criss-cross ridges around the edge of something so that it is easier to grip

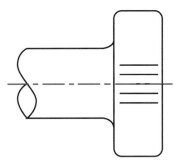

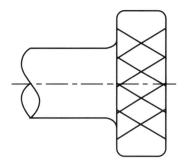

(a) Straight knurling

(b) Diamond knurling
(drawn at 30 degrees)

Figure 3.25 Knurl types

Check your understanding

6 What is the meaning of tolerance?
7 What is a leader line used for?
8 What is the meaning of the abbreviation AF used on an engineering drawing?
9 What is a countersink?
10 Name **two** types of knurling.

Now test yourself

TESTED ⬤

1 Use the internet to find examples of the following engineering drawings:
   + Orthographic drawing
   + Assembly drawing
   + Flowchart
2 On paper, draw each of the following line types:
   + Hidden detail line
   + Dimension line
   + Centre line
   + Break line

3.3.1 Advantages and disadvantages of using CAD drawing software versus manual drawing techniques

A designer has to consider several factors when deciding which drawing type or method to use, including the:

+ type of drawing that is required, for example quick sketch and component drawing
+ tools and equipment they have access to.

There are advantages and disadvantages to both manual drawing and computer-aided design (CAD), as described in Table 3.5.

> **Computer-aided design (CAD)** Computer software used to develop designs for new products or components

Table 3.5 Advantages and disadvantages of manual drawing and CAD

Drawing type	Advantages	Disadvantages
Manual	Low-cost drawing equipment	Takes a long time to draw
	Secure storage of drawings	Paper degrades over time
	Short training time	Cannot link to computer numerically controlled (CNC) machines
CAD	Better quality designs	High cost of set up
	Easy saving and sharing of files	Computer subject to hacking and viruses
	Ability to create 3D models	Long training time
	Easy data storage and accessibility	
	Faster to modify and reproduce drawings	
	Can link to computer numerically controlled (CNC) machines	

Exam tip

Examiners may ask you to describe the advantages and disadvantages of using CAD software. Try to remember at least two advantages and two disadvantages for both manual drawing and CAD.

Typical mistake

A common mistake is to write very generally about the advantages and disadvantages of manual drawing and CAD. You will gain more marks if you are specific about each one. For example, an advantage of manual drawing is 'low cost of drawing equipment' – do not just write 'low cost'.

Check your understanding

11 Describe **one** disadvantage of manual drawing compared to CAD.
12 State **two** advantages of using CAD software.
13 Describe **one** disadvantage of using CAD software.

Now test yourself

TESTED

Give **two** advantages and **one** disadvantage of using manual drawings.

1 Which of the following drawing types is shown in Figure 3.26. [1]

  a Oblique ☐

  b Isometric ☐

  c Orthographic ☐

  d Assembly ☐

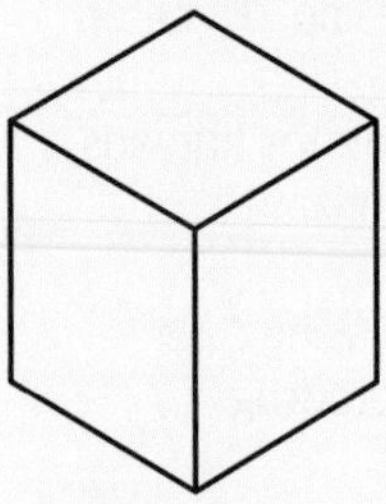

Figure 3.26

2 Identify the **three** main elevation views on an orthographic drawing. [3]

3 What angles do the following drawing types use?

  a Oblique ☐ [1]

  b Isometric ☐ [1]

4 State **one** advantage of exploded views. [1]

5 Explain why assembly drawings are used in engineering. [3]

6 Identify the diagram type shown in Figure 3.27. [1]

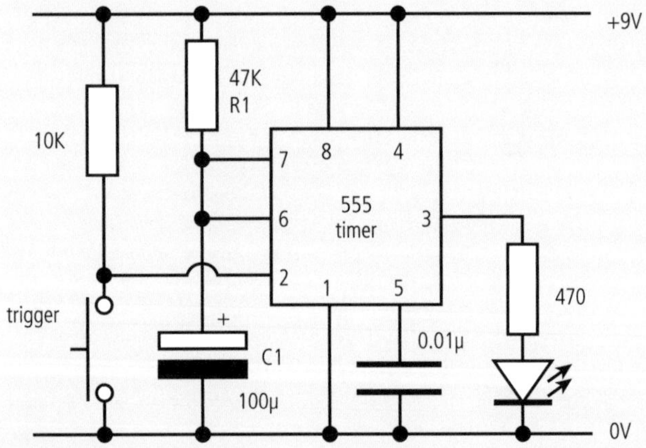

Figure 3.27

7 Discuss why standard conventions, such as BS 8888, are used by designers and manufacturers. [4]

8 Identify what the symbol shown in Figure 3.28 represents on an engineering drawing. [2]

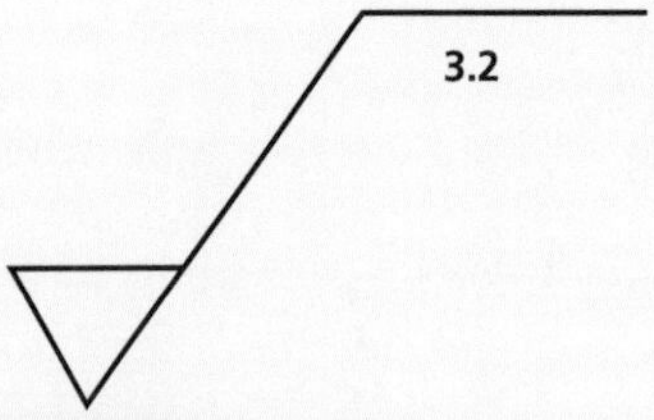

3.2

Figure 3.28

9 State the meaning of the abbreviation SQ used on an engineering drawing. [1]

10 Choose **one** of the following that represents the mechanical feature shown in Figure 3.29. [1]

  a External thread ☐

  b Internal thread ☐

  c Straight knurling ☐

  d Diamond knurling ☐

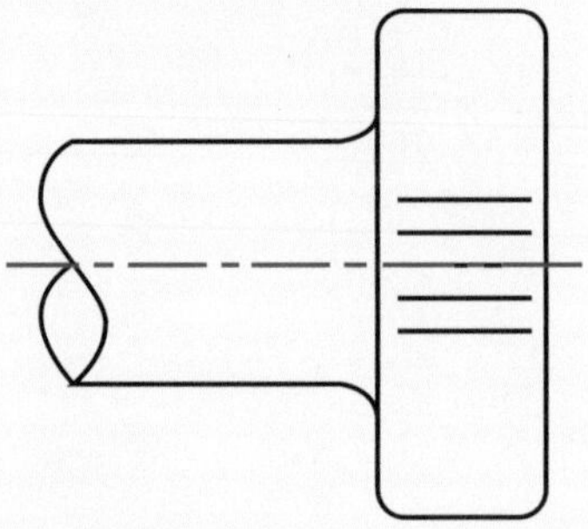

Figure 3.29

Exam checklist

In this topic, you learned about the following:

+ Types of drawing used in engineering and the typical applications and relative advantages and disadvantages of each drawing technique:
    + Freehand sketching
    + Isometric drawing
    + Oblique drawing
    + Orthographic drawing
    + Exploded views
    + Assembly drawings
    + Block diagrams
    + Flowcharts
    + Circuit diagrams
    + Wiring diagrams
+ Working drawings:
    + 2D engineering drawings using third angle orthographic projection
+ Standard conventions in BS 8888 and how these are applied (title block, metric units of measurement, scale, tolerance)
+ Standard conventions for dimensions (linear measurements, radius, diameter, surface finish)
+ Meaning of line types (outline, hidden detail, centre line, projection, dimension, leader line)
+ Abbreviations (across flats, centre line, diameter, drawing, material, square)
+ Representations of mechanical features (threads, holes, chamfers, countersinks, knurls)
+ Using CAD drawing software:
    + Advantages and disadvantages of using CAD drawing software versus manual drawing techniques.

Topic area 4: Evaluating design ideas

4.1 Methods of evaluating design ideas

Designers evaluate design ideas in order to get feedback and make improvements. They will use a range of methods to help them choose the best idea, for example:
+ production of models
+ qualitative comparison with the design brief and specification
+ ranking matrices
+ quality function deployment (QFD).

4.1.1 Production of models

A designer will make a model to give a client a complete idea of how a product would look in real life. A model can help to:
+ clarify the product's purpose, function and appearance
+ enable designers and engineers to better understand the product prior to manufacture.

A prototype will then be made to fully test the functionality of the product.

4.1.2 Qualitative comparison with the design brief and specification

After a designer has made a model, it can be compared against the original requirements of the design brief and the design specification. This is to ensure that:
+ the client's requirements are met
+ the product's aesthetics are correct
+ there are no errors within the design.

4.1.3 Ranking matrices

Ranking matrices (also called decision matrices) are used to help make decisions when comparing products when there is more than one option and several factors to consider.

> **Ranking matrices** Tables with numbers to support decision making when comparing products when there is more than one option and several factors to consider

Simple rank matrix

The simple rank matrix in Table 4.1 can be used to help decide which type of electric bike to buy:
+ The criteria against which the scores will be evaluated are listed in the first column.
+ The options (electric bike A, B or C) are then listed in columns 2–4.
+ A score is then allocated to each criterion for each option. In this instance, a scale of 1–5 is used, with 1 being a poor score and 5 a high score.
+ The totals are then added up, and the option with the highest score is the winner. (However, electric bikes B and C both score the same in this example).

Table 4.1 Simple rank matrix for three electric bikes (A, B and C)

Criteria	Options		
	Electric bike A	Electric bike B	Electric bike C
Cost	5	3	2
Performance	2	5	4
Reliability	4	2	5
Practicality	1	5	4
Battery life	4	3	3
Total	**16**	**18**	**18**

Weighted rank matrix

A weighted rank matrix helps to give weight (importance) to criteria, in order to make the decision process more specific.

As per the simple rank matrix in Table 4.1, the criteria are still given a score from 1 to 5 (with 1 being a poor score and 5 a high score), but each score is then multiplied by the criteria weighting and added together to give a total. As you can see in Table 4.2, this now makes the decision process clearer – and electric bike B is the winner with the highest score.

Table 4.2 Weighted rank matrix for three electric bikes (A, B and C)

Criteria	Weighting	Options					
		Electric bike A		Electric bike B		Electric bike C	
		Score	Total	Score	Total	Score	Total
Cost	5	5	25	3	15	2	10
Performance	3	2	6	5	15	4	12
Reliability	1	4	4	2	2	5	5
Practicality	2	1	2	5	10	4	8
Battery life	4	4	16	3	12	3	12
Total			**53**		**54**		**47**

Datum rank matrix

A datum rank matrix uses an existing product to make a comparison with products being developed.

> **Datum** A fixed reference point

In Table 4.3, electric bike D (datum) is being compared with options A to C:
+ If a criterion is better, it is given a +.
+ If worse, it is given a –.
+ If the same, it is given an S.

The totals are then added up, with a score of 1 for a +, a score of –1 for a –, and a score of zero for an S.

As you can see from Table 4.3, electric bike C has the highest score when compared against the datum, electric bike D. This method of evaluation can help a designer make decisions when developing new products compared to existing ones.

Table 4.3 Datum rank matrix for four electric bikes (A, B, C and D)

Criteria	Options			
	Electric bike A	Electric bike B	Electric bike C	Electric bike D
Cost	S	–	–	DATUM
Performance	–	+	+	
Reliability	+	+	+	
Practicality	–	S	+	
Battery life	S	–	–	
+ Better	1	2	3	
– Worse	2	2	2	
Same	$2 \times 0 = 0$	$1 \times 0 = 0$	$0 \times 0 = 0$	
Total	**−1**	**0**	**1**	

4.1.4 Quality function deployment (QFD)

Quality function deployment (QFD) is a more complex ranking matrix. It is an evaluation tool that allows the designer to focus on what is important in a design, in order to satisfy the customer:

✚ It assesses the customer's qualitative needs and wants and converts them into quantitative criteria.
✚ This then allows a detailed design specification and product design to be produced.
✚ The quantitative criteria can then be used to evaluate the product.

QFDs can also be used to plan large engineering projects and can be very complex.

> **Typical mistake**
>
> Do not confuse the qualitative comparison with the design brief and specification with a quantitative comparison. This is a quality check to make sure the client is happy and the product is correct with no errors.

> **Quality function deployment (QFD)**
> Evaluation tool that helps transform the customer's needs and wants into a product design

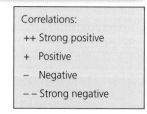

Correlations:
++ Strong positive
+ Positive
– Negative
–– Strong negative

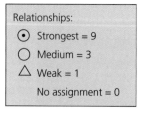

Relationships:
⊙ Strongest = 9
○ Medium = 3
△ Weak = 1
No assignment = 0

Competitors:
A Phone A (*)
B Phone B
C Phone C

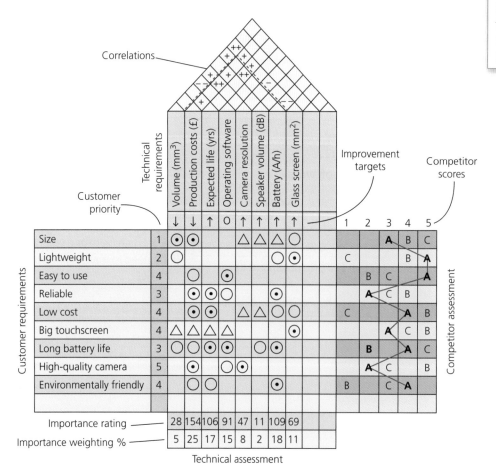

Figure 4.1 Quality function deployment (QFD) for a mobile phone

Check your understanding

1 Describe why a designer compares a model against the design brief.
2 How does a weighted rank matrix make the decision process more specific?
3 A designer may compare a model against the design brief and use a ranking matrix to evaluate a design idea. Identify **one** other method to evaluate a design idea.

Now test yourself TESTED

1 List the different ways in which QFD may help a designer.
2 Choose **two** similar products – for example types of smartwatch, hair dryers, Bluetooth headphones – and suggest which is better:
   + Describe the design features and details that in your opinion make one product better than the other. There is no need to comment on the brand.
   + List the main functions of each product.

4.2 Modelling methods

REVISED

Designers use a wide range of modelling methods to create virtual and physical models. This allows them to visualise and test how a product looks and performs in 3D. It is also an excellent way of checking a product's potential for success.

Modelling methods include:
+ virtual (3D CAD)
+ card
+ block
+ breadboarding
+ 3D printing.

4.2.1 Virtual modelling using 3D CAD

Virtual modelling uses computer-aided design (CAD) software.

Table 4.4 Virtual modelling using 3D CAD

Equipment	Computer mouse, CAD software
Stages	Objects are drawn in 2D then converted to 3D.
Benefits	3D CAD provides better visualisation of a product and all its component parts.
	The components can be animated and their function observed.

Figure 4.2 Virtual modelling using 3D CAD

4.2.2 Modelling using card

Card and cardboard sheet allow modelling ideas to be created quickly.

Table 4.5 Modelling using card

Equipment	Card, scissors, craft knife, cutting mat, glue
Stages	A template is drawn, cut out and glued together.
Benefits	It provides a 3D version of a product.
	It can be used to test some functionality.

Figure 4.3 Card modelling

4.2.3 Modelling using block foam

Block foam also allows modelling ideas to be created quickly.

Table 4.6 Modelling using block foam

Equipment	Block foam, glue, hand tools, files, hot cutter
Stages	A design template is drawn and glued to the foam.
	The design is cut out using hand tools and files or a hot cutter.
	The foam is then sanded to achieve a smooth surface.
	A finish is added.
Benefits	It provides a 3D version of a product.
	It can be used to test some functionality.

Figure 4.4 Foam modelling

4.2.4 Breadboarding

Breadboarding involves developing and testing electrical circuit designs on a solderless construction base known as a breadboard.

Breadboard A solderless construction base used for prototyping electronics

Table 4.7 Breadboarding

Equipment	Breadboard, power supply, electronic components, jump wires
Stages	A circuit diagram is used and all the components are selected.
	The components are then placed onto the breadboard.
Benefit	The components can be easily swapped to improve or fix the circuit.

Figure 4.5 Breadboarding

Check your understanding and progress at **www.hoddereducation.co.uk/myrevisionnotes**

4.2.5 3D printing

3D printing is an additive manufacturing process that creates a physical object from a digital 3D CAD model.

Table 4.8 3D printing

Equipment	3D printer, extrusion material, finishing materials
Stages	Create a 3D model in CAD and convert the CAD file to an STL file.
	Send the STL file to the 3D printer and select 'Print' to start the build.
	The 3D printer builds the 3D object, layer by layer, as it extrudes material from the nozzle onto the print bed.
	Remove the 3D model from the print bed when cool and remove any supports from the 3D model.
	Clean up the surface of the 3D model and add a finish.
Benefit	From the 3D CAD model, complex shapes can be produced.

Additive manufacturing
Process by which layers of material are added to create a 3D object

STL Abbreviation for a stereolithography file used in CAD software systems to create 3D objects

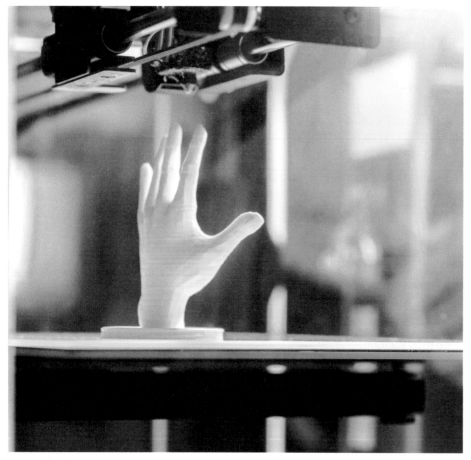

Figure 4.6 3D printing

> **Exam tip**
>
> An exam question may ask what tools are *most* suitable for a particular modelling method. All options may be used, but only one is *most* suitable. Make sure you read all the options in the exam – not just the first one.

Topic area 4: Evaluating design ideas

4.2.6 Advantages and disadvantages of each modelling method

The advantages and disadvantages of the various modelling methods are summarised in Table 4.9.

Table 4.9 Advantages and disadvantages of different modelling methods

Modelling method	Advantages	Disadvantages
Virtual (3D CAD)	Very accurate modelling No need to make a physical model	Requires skill training to use the software High initial set-up costs
Card	Low-cost materials Limited equipment needs	Cannot be used for full testing, as materials differ from the made product
Block foam	Light and easy to cut Can produce an excellent finish	Requires patience to use, as it can crumble at the edges Cannot be used for full testing, as materials differ from the made product
Breadboarding	No soldering required Easy to check a circuit works	Only a few components can be tested at a time Electronics knowledge is required
3D printing	A very accurate 3D model is made Uses a 3D CAD model	Equipment can be expensive to buy Limited materials available Takes a long time to make a model

Exam tip

An exam question may ask you to describe the advantages and disadvantages of modelling methods. Aim to remember at least one advantage and one disadvantage for each modelling method.

Check your understanding

4 Name **two** items of equipment needed when block foam modelling.
5 Describe, step by step, how a block foam model is made.
6 What is meant by breadboarding?
7 What is meant by the term additive manufacturing?

Now test yourself

TESTED ◯

List the **five** stages of 3D printing.

4.3 Methods of evaluating a design outcome

REVISED ◯

When evaluating a design outcome, a designer will use a wide range of methods, including:
+ measuring the dimensions of a product
+ measuring the functionality of a product
+ quantitative comparison with the design brief and specification
+ user testing.

Exam tip

Examiners may ask questions about the methods of evaluating a design outcome. Make sure you know the four methods of measuring the dimensions and functionality, carrying out quantitative comparison with the design brief and specification, and user testing.

4.3.1 Measuring the dimensions of a product

A designer will use various measuring tools to accurately check the dimensions of a product. Typical tools are listed in Table 4.10.

Table 4.10 Measurements and measuring tools

Measurement	Measuring tool	Advantages	Disadvantages
Linear dimensions	Steel rule (sometimes called an engineer's rule)	Easy to use and low cost	Can only measure linear dimensions with an accuracy of 0.5 mm up to distances of 1 m, i.e. lengths, widths and depths of an object
	Tape measure	Available in a variety of lengths from 2 to 100 m	Can only measure linear dimensions with an accuracy of 0.5 mm
	Vernier calliper	Used for very accurate measurements, typically up to ±0.01 mm Can measure internal, external and depth measurements	Skill is required to use correctly More costly than a steel rule
Circular dimensions	Micrometer	Used for precision measuring with an accuracy of ±0.001 mm Suitable to measure round objects Available to measure both external or internal diameters	Limited maximum measuring range Skill is required to use correctly Costly and easily damaged if dropped or used incorrectly
Weight	Weight scales	Used for accurate measuring of weight up to 5 kg in 1 g increments Easy to read and adjustable	Large or irregular-shaped items over 5 kg cannot be weighed
Electronic circuits	Multimeter	Accurate readings of voltage, current and resistance Easy to read and select settings	Skill and electronics knowledge required to operate

4.3.2 Measuring the functionality of a product

When evaluating functionality, designers will address the following issues:
+ Does the product do what it is designed to do?
+ Is the product easy to use?
+ How well does the product perform?
+ How successful are the product's features and details?

4.3.3 Quantitative comparison with the design brief and specification

After a designer has made a final product, it can be compared against the original requirements of the design brief and design specification. This is to ensure that the:
+ client's technical requirements have been met
+ product's dimensions are correct
+ product functions as required.

4.3.4 User testing

User testing is carried out by people with no connection to the designer, so their views are unbiased. This helps to:
+ provide valuable feedback about a product
+ identify unforeseen errors or missed opportunities
+ allow errors to be corrected before manufacturing begins.

Figure 4.7 User testing allows errors to be corrected before manufacturing begins

4.3.5 Advantages and disadvantages of each method of evaluating a design outcome

There are advantages and disadvantages of the various evaluation methods, as shown in Table 4.11.

Table 4.11 Advantages and disadvantages of evaluation methods

Evaluation method	Advantages	Disadvantages
Measuring the dimensions of a product	Accurate measuring ensures correct dimensions are achieved	Using some measuring tools accurately requires training
Measuring the functionality of a product	Checks that a product functions correctly before manufacture	Takes time to thoroughly test the function of a product
Quantitative comparison with the design brief and specification	Ensures client requirements have been met	Takes time to check the design brief and specification
User testing	Provides feedback to improve the design Low cost	Must ensure the users' views are unbiased

Check your understanding and progress at **www.hoddereducation.co.uk/myrevisionnotes**

4.3.6 Reasons for identifying potential modifications and improvements to the design

After reviewing all the feedback from the evaluation, a designer will identify potential areas for development and give reasons to modify and improve the product. This may include the following:

+ Product operation – Was it easy to use? Was it ergonomic? If not, why not?
+ Product scale – Was it the correct size and weight? If not, why not?
+ Product function – Did it do the job it was designed for? If not, why not?

> **Typical mistake**
>
> When evaluating a model, do not confuse a quantitative comparison of the design brief and specification with a qualitative comparison. The dimensions of the product are measured and the function is checked.

> **Exam tip**
>
> Examiners may ask you to describe the advantages and disadvantages of evaluation methods. Aim to remember at least one advantage and one disadvantage for each evaluation method.

Check your understanding

8 Describe one disadvantage of user testing.

9 State **one** tool used to measure linear dimensions.

10 Describe **one** disadvantage of using tools to measure linear dimensions.

11 State **one** tool used to measure electronic circuits.

12 Why is user testing often used by designers to evaluate products?

Now test yourself

TESTED ⬤

Creating a new product is a complex process and designers will take time to fully evaluate the final product against the original design brief and engineering specification. After they have carried out all the quantitative checks, they will carry out some qualitative checks.

1 Why is it important for designers to carry out user testing on the final product?

2 Consider the following requirements for a pair of Bluetooth headphones and decide which are quantitative and which are qualitative:
   + long battery life
   + CE marked
   + soft earpieces
   + comfortable to wear and operate
   + dimensions of earpieces 90 60 20 mm
   + smooth textured head strap.

Exam-style questions

1 Describe why a design engineer evaluates the product specification. [2]

2 Describe **one** way that quality function deployment (QFD) may help a product designer. [2]

3 State **one** advantage of using virtual 3D CAD. [1]

4 Identify **two** advantages of producing a model using card. [2]

5 State **one** disadvantage of breadboarding. [1]

6 Identify the type of manufacturing process that 3D printing uses. [1]

7 State **one** method used to evaluate a design outcome. [1]

8 State **two** measuring tools that can be used to measure the dimensions of a product. [2]

9 Identify **one** tool used to measure the weight of a product [1]

10 Explain why a designer may need to modify the design of a new product once it has been evaluated. [4]

Exam checklist

In this topic, you learned about the following:
+ Methods of evaluating design ideas:
    + Production of models
    + Qualitative comparison with the design brief and specification
    + Ranking matrices
    + Quality function deployment (QFD)
+ Modelling methods:
    + Virtual (3D CAD)
    + Card
    + Block foam
    + Breadboarding
    + 3D printing
+ Information that can be obtained using each method
+ Equipment required and stages involved in each method
+ Advantages and disadvantages of each method
+ Methods of evaluating a design outcome:
    + Measuring the dimensions of a product
    + Measuring the functionality of a product
    + Quantitative comparison with the design brief and specification
    + User testing
    + Advantages and disadvantages of each method
    + Reasons for identifying potential modifications and improvements to the design.

Check your understanding and progress at **www.hoddereducation.co.uk/myrevisionnotes**

Answers to 'Check your understanding' questions

Topic area 1 Designing processes

1 The various stages of a product's evolution, from its beginning to its end
2 Development of a product with a clear understanding of user needs and requirements
3 Any two of the following: Reduce; Reuse; Recycle; Rethink; Refuse; Repair
4 To ensure products are comfortable to use, are easy to understand and fit the user they are designed for
5 Relies on good user feedback, which may be limited depending on who is chosen as a user
6 Any of the following:
   + ensures products cause minimal harm to people and the planet
   + uses the 6 Rs of sustainability
7 Any two of the following:
   + what product needs to be designed
   + what the product will be used for
   + who the product is for
   + where the product will be sold
   + where the product will be used
   + when the prototype must be finished
8 Gathering original information first hand
9 Cost or Customer
10 Size or Safety
11 To create a 3D model that demonstrates the functionality of a product
12 Any of the following:
   + no need to make a physical model
   + very accurate modelling
13 Any of the following:
   + can be time-consuming
   + a level of skill required
   + can be costly
   + model might not look like the final product

Topic area 2 Design requirements

1 Wants are characteristics of a product that are desirable but not essential.
2 Criteria that can be expressed as a number and quantified by hard facts
3 Objective criteria are based on facts, reliable and not influenced by personal feelings or opinions; subjective criteria are based on personal feelings, tastes or opinions.
4 Size
5 Any of the following:
   + Where will the product be used?
   + Will the product perform its intended task?
   + What should be tested to check the product's performance?
   + What maintenance will be needed during use?
6 Any of the following:
   + one-off production
   + batch production
   + mass production
7 Any two of the following:
   + tube: hexagon; circle; square; rectangle
   + solid: plate; square; hexagon; round; angle; channel
8 a Forming
   b Shaping
9 Laws proposed by the government and made official by Acts of Parliament.
10 When new technology is created because of research and development (R&D), resulting in new products that are 'pushed' into the market, with or without demand
11 Guidelines on how to meet legislation
12 A policy of producing consumer goods that rapidly become obsolete (out of date and therefore unusable), either due to changes in design or unavailability of spare parts
13

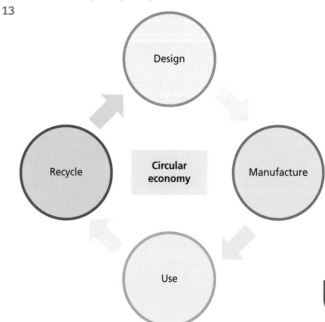

73

Topic area 3 Communicating design outcomes

1 A drawing of a system or product where separate parts are represented by blocks and connected by lines showing their relationship to one another.

2 To show how components fit together or the order of assembly

3 To show how various processes are linked together to achieve a specific outcome

4 Any of the following:
   + enables an idea to be visualised quickly
   + uses a universal language that all designers understand

5 Any of the following:
   + difficult to dimension
   + looks less realistic than isometric drawing
   + cannot see the rear of the object

6 The acceptable amount of variation allowed on a dimension

7 To connect a graphical representation on the drawing to some text

8 Across flats

9 A conical hole cut into a material at a given angle so that a bolt or screw can be sunk below the surface

10 Straight and diamond

11 Any of the following:
   + takes a long time to draw
   + paper degrades over time
   + cannot link to computer numerically controlled (CNC) machines

12 Any two of the following:
   + better quality designs
   + easy saving and sharing of files
   + ability to create 3D models
   + easy data storage and accessibility
   + faster to modify and reproduce drawings
   + can link to computer numerically controlled (CNC) machines

13 Any of the following:
   + high cost of set up
   + computer subject to hacking and viruses
   + long training time

Topic area 4 Evaluating design ideas

1 To check:
   + the client's requirements are met
   + the product's aesthetics are correct
   + there are no errors within the design

2 By giving weight (importance) to the criteria

3 Use a quality function deployment (QFD)

4 Any two of the following:
   + block foam
   + glue
   + hand tools
   + files
   + hot cutter

5 Stages:
   + design template drawn and glued to the foam
   + design cut out using hand tools and files or a hot cutter
   + foam sanded to achieve a smooth surface
   + finish added

6 Developing and testing electrical circuit designs on a solderless construction base known as a breadboard

7 Process by which layers of material are added to create a 3D object

8 Must ensure the users' views are unbiased

9 Any of the following:
   + steel rule
   + tape measure
   + vernier calliper

10 Any of the following:
   + steel rules and tape measures only have an accuracy of 0.5 mm
   + skill required to use correctly

11 Multimeter

12 User testing can:
   + provide valuable feedback about a product
   + identify unforeseen errors or missed opportunities
   + allow errors to be corrected before manufacturing begins.

Answers to exam-style questions

Topic area 1 Designing processes

1 a Computer-gaming chair [1 mark]
2 Any of the following for 1 mark:
   + focuses on users with specific needs
   + enables a wide range of users to participate in everyday activities
3 Requires the attitudes of people and industry to change for it to work [1 mark]
4 Aesthetics [1 mark]
5 + Description: where information is gathered from sources that already exist [1 mark]
   + Example: choose from internet, TV programmes, databases, textbooks, newspapers [1 mark]
6 Design [1 mark], make [1 mark], evaluate [1 mark]
7 1 mark for each reason and 1 mark for each description, to a maximum of 4 marks:
   + To test proportions [1 mark], to ensure the relationship between the size of different parts of a product is correct or attractive [1 mark]
   + To test scale [1 mark], to ensure overall dimensions of the product are correct or attractive [1 mark]
   + To test function [1 mark], to ensure the product operates in a proper or particular way [1 mark]
8 1 mark for an example and 1 mark for each reason, to a maximum of 4 marks:
   + Example: the use of plastic bottles [1 mark]
   + Designers look at how a product will affect the environment, for example plastics cause pollution [1 mark].
   + Sustainable design can help cut down on the number of plastics used [1 mark].
   + This ensures minimal or no harm to people/the planet [1 mark] through the use of the 6 Rs of sustainability [1 mark].
9 Up to 6 marks for a discussion or detailed explanation of the advantages and disadvantages of using the linear design strategy when designing a lifejacket for a water sports activity:
   + Level 3 (5–6 marks): detailed discussion including a variety of advantages and disadvantages of using the linear design strategy when designing a lifejacket for a water sports activity, showing understanding of all the points stated.
   + Level 2 (3–4 marks): adequate discussion including a few advantages and disadvantages of using the linear design strategy when designing a lifejacket for a water sports activity, showing understanding of all the points stated.
   + Level 1 (1–2 marks): basic discussion including only either advantages or disadvantages of using the linear design strategy when designing a

lifejacket for a water sports activity, showing understanding of all the points stated.
Responses may include reference to the following:
Advantages:
   + Error catching is early in the design process.
   + Risk is managed throughout the process using strict controls.
   + All safety features of the lifejacket are thoroughly tested.
Disadvantages:
   + Each phase must be completed before going on to the next.
   + Research may focus on general water sports use only.
   + There is no flexibility to change or improve the design.
   + There is no customer feedback used to alter the design.
   + Projects may take a long time to complete.
   + Projects can be expensive.
10 Up to 6 marks for a discussion or detailed explanation of the advantages and disadvantages of using the sustainable design strategy when designing an electric car:
   + Level 3 (5–6 marks): detailed discussion including a variety of advantages and disadvantages of using the sustainable design strategy when designing an electric car, showing understanding of all the points stated.
   + Level 2 (3–4 marks): adequate discussion including a few advantages and disadvantages of using the sustainable design strategy when designing an electric car, showing understanding of all the points stated.
   + Level 1 (1–2 marks): basic discussion including only either advantages or disadvantages of using the sustainable design strategy when designing an electric car, showing understanding of all the points stated.
Responses may include reference to:
Advantages:
   + It reduces the impact on the environment by reducing pollution and material waste.
   + It ensures a better environmental quality for present and future generations.
   + The 6 Rs model helps with improved management of materials.
Disadvantages:
   + Costs can be high with the use of new technologies.
   + The 6 Rs model can be restrictive on material availability.
   + It takes time and effort to ensure products are sustainable.

Topic area 2 Design requirements

1   A characteristic that a product must have [1 mark]
2   b   Numbers and statistics [1 mark]
3   1 mark for any of the following:
    + How much does it cost to manufacture the product?
    + Will the product be affordable?
    + How much will a customer pay for the product?
    + Will the product be profitable?
4   1 mark for any of the following:
    + painting
    + lacquering
    + electroplating
    + ceramic coating
    + powder coating
5   1 mark for any of the following:
    + cost of labour to make the product
    + cost of equipment and processes
    + cost of materials
    + cost of factory overheads, such as rent, heating, insurance and power
6   1 mark for an influence and 1 mark for each explanation, to a maximum of 4 marks:
    + what market there is for a new product [1 mark] by carrying out market research [1 mark]
    + which new technologies are available [1 mark] to ensure a product is fully up to date [1 mark]
    + which standards and legislation must be followed [1 mark] to ensure laws are not broken [1 mark]
    + how a product can be made sustainable [1 mark] by checking the materials being used/how to dispose of them [1 mark]
7   1 mark for each relevant point, to a maximum of 3 marks:
    + Legislation ensures products comply with relevant safety standards [1 mark].
    + A designer could face legal action if a product was found to be unsafe or have caused harm [1 mark].
    + Users may take legal action against a company [1 mark].
    + If a company or an individual is shown to have broken the law, they can be prosecuted [1 mark].
8   3 marks for the description and 1 mark for the example:
    + Planned obsolescence is a policy of producing consumer goods that rapidly become obsolete/out of date [1 mark] and therefore unusable [1 mark], either due to changes in design or unavailability of spare parts [1 mark].
    + An example would be using irreplaceable batteries in tech products [1 mark].
9   1 mark for a correct identification and 1 mark for a linked description – any of the following:
    + Reduce [1 mark]: cut down on the amount of material used as much as possible [1 mark].
    + Recycle [1 mark]: reprocess a material or product and make something else [1 mark].
    + Reuse [1 mark]: use a product to make something else with all or parts of it [1 mark].
    + Refuse [1 mark]: do not use a material if it is bad for the environment [1 mark].
    + Rethink [1 mark]: design in a way that considers people and the environment [1 mark].
    + Repair [1 mark]: when a product breaks down or stops working properly, fix it [1 mark].
10 A system designed to reduce the impact of the use of materials on the environment [1 mark] by reducing waste and pollution [1 mark] and by reusing and recycling materials [1 mark]

Topic area 3 Communicating design outcomes

1   b   Isometric [1 mark]
2   Front [1 mark], side [1 mark] and plan/top [1 mark]
3   a   45 degrees [1 mark]
    b   30 degrees [1 mark]
4   They help understanding of how parts are arranged [1 mark].
5   1 mark for each point, to a maximum of 3 marks:
    + Assembly drawings give the big picture of a completed project [1 mark].
    + They show how separate components are joined together [1 mark] to make a whole product [1 mark].
    + They show the general arrangement of the parts [1 mark].
    + They provide a parts list and parts numbers [1 mark].
6   Circuit diagram [1 mark]
7   1 mark for each relevant point, to a maximum of 4 marks. For example:
    + understanding of the same conventions [1 mark]
    + use of the same symbols and terminology [1 mark]
    + reduction in errors from misunderstanding [1 mark]
    + reduction in costs due to less scrap and rework [1 mark]
    + improved productivity [1 mark]
    + better product quality [1 mark]
    + BS 8888 is a reference source for engineering drawings [1 mark]
8   Surface finish [1 mark] of 3.2 µm Ra [1 mark]
9   Square [1 mark]
10 c   Straight knurling [1 mark]

Topic area 4 Evaluating design ideas

1 1 mark for each relevant point, to a maximum of 2 marks:
+ provides useful evaluation criteria of what was initially required for the product [1 mark]
+ to make sure the client's requirements are met [1 mark]

2 QFD allows the designer to focus on what is important in a design to satisfy the customer [1 mark]. It is also used to plan engineering projects and can be very complex [1 mark].

3 1 mark for any of the following:
+ very accurate modelling
+ no need to make a physical model

4 1 mark for each advantage, to a maximum of 2 marks:
+ provides a 3D version of a product
+ can be used to test some functionality
+ limited equipment needs
+ low-cost materials

5 1 mark for any of the following:
+ only a few components can be tested at a time
+ electronics knowledge is required

6 Additive manufacturing [1 mark]

7 1 mark for any of the following:
+ measuring the dimensions of a product
+ measuring the functionality of a product
+ quantitative comparison with the design brief and specification
+ user testing

8 1 mark for each method, to a maximum of 2 marks:
+ steel rule
+ tape measure
+ vernier calliper
+ micrometer

9 1 mark for any of the following:
+ weight scales

10 1 mark for each relevant point, to a maximum of 4 marks:
+ to identify potential areas for development
+ to ensure the product operates correctly
+ to check that the product is easy to use
+ to ensure the product is ergonomic
+ to check the scale of the product is correct
+ to check the size of the product is correct
+ to check the weight of the product is correct
+ to ensure the product does the job it was designed for

Glossary

Additive manufacturing Process by which layers of material are added to create a 3D object. **67**

Additive Where material is added to create a product, for example 3D printing. **34**

Aesthetics How a product appeals to the senses; something that is pleasing in appearance based on its form, shape, symmetry, texture, colour and proportion. **19**

Anthropometric Relating to the study of measurements of the human body. **15**

Assembly drawing Drawing that shows how separate components are joined together to make a whole product. **45**

Assembly Process of fitting component parts together to make a whole product. **34**

Automate To make something operate by automatic CNC machinery or equipment. **27**

Blind holes Holes that do not break through to the other side of the workpiece. **56**

Block diagrams Drawings of systems or products where separate parts are represented by blocks and connected by lines showing their relationship to one another. **45**

Brazing Joining metal parts by melting and flowing a filler metal into the joint, using heat provided by either a flame or an oven. **33**

Breadboard A solderless construction base used for prototyping electronics. **66**

Capital costs Fixed, one-time expenses used in the production of goods to purchase buildings, machinery and equipment. **26**

Carbon dioxide Colourless and non-flammable gas that exists in the Earth's atmosphere; rising carbon dioxide levels are contributing to global warming. **39**

Ceramics Materials that are typically an oxide, a nitride or a carbide of a metal. **30**

Chamfer A transitional edge (or cut-away) between two faces of an object. **56**

Circuit diagrams Conventional graphical representations of electrical circuits, where separate electrical components are connected to one another. **46**

Circular economy System designed to reduce the impact of the use of materials on the environment by reducing waste and pollution and reusing and recycling materials. **39**

Composites Materials consisting of two (or more) different materials bonded together. **32**

Compressive forces Pressing or squeezing forces. **32**

Computer-aided design (CAD) Using computer software to develop designs for new products or components. **20, 58**

Cotter pins Pin-shaped pieces of metal that pass through a hole to fasten two parts of a mechanism together. **33**

Countersinks Conical holes cut into an object at a given angle so that a bolt or screw can be sunk below the surface. **56**

Datum A fixed reference point. **62**

Design brief The context for a design problem, with reference to user needs and wants, product performance, end use, scale of production, time limits and target market. **17**

Diameter A straight line from one side of a circle to the other through the centre. **52**

Die Tool that imparts a desired shape on a metal or polymer; typically a perforated block through which a metal or polymer is forced by extrusion. **32**

Dimensioned Marked with numerical values to communicate the sizes of key features (measurements are usually in millimetres). **49**

Disassembly Taking something apart, for example a product or piece of equipment. **19**

Electroplating Process that produces a metal coating on a part by placing it in a chemical bath and passing an electrical current through it. **34**

Elevations Viewing projections within an orthographic drawing, i.e. front, side and plan. **44**

Engineering design specification Detailed document that defines the criteria required for a new product. **19**

Ergonomic design Development of products using anthropometric data so that they perfectly fit the people who use them. **15**

Exploded view Drawing where the components of a product are drawn slightly separated from each other and suspended in space, to show how they fit together or the order of assembly. **44**

Extrusion Forming process in which a metal or polymer is forced through a die. **31**

Finishing Process that changes the surface of a material in a useful way. **33**

Firing Process that turns ceramic powders into solid objects using high temperatures. **30**

Flowcharts Block diagrams that show how various processes are linked together to achieve a specific outcome. **46**

Check your understanding and progress at **www.hoddereducation.co.uk/myrevisionnotes**

Forming Process that changes the shape of a material without changing its state. **31**

Freehand sketching Drawing without the use of measuring instruments. **42**

Inclusive design Development of a product so that it can be used by as many people as possible, regardless of their age, ability or background. **12**

Isometric drawing A 3D pictorial drawing that focuses on the edge of an object and uses an angle of 30 degrees to the horizontal. **43**

Iterative design Development of a product through modelling and repeated testing. **12**

Joining Process that attaches separate pieces of material together to make a product. **32**

Knurling Process of machining a series of straight or criss-cross ridges around the edge of something so that it is easier to grip. **57**

Lathe Machine that rotates a workpiece about an axis to perform various operations such as cutting, drilling, boring, facing and turning. **29**

Leader line terminator The type of shape used to end a line. **54**

Legislation Laws proposed by the government and made official by Acts of Parliament. **38**

Limited variation Where there is little change in the design of a product. **27**

Linear design Development of a product through a series of sequential stages. **11**

Manual dexterity The skill of using the hands to carry out a task with precision. **11**

Market pull When a need for a product arises from customer demand, which 'pulls' development of the product. **36**

Market research Process of gathering information about the needs and preferences of potential customers. **18**

Melting point Temperature at which a substance changes from a solid to a liquid. **31**

Metals Materials typically made by processing an ore that has been mined or quarried. **30**

Models Virtual or physical 3D objects that demonstrate the aesthetics of products. **20**

Need Critical aspect of a product that makes it fit for purpose. **23**

Objective Based on facts and reliable; not influenced by personal feelings or opinions. **23**

Oblique drawing A 3D pictorial drawing that focuses on the face of an object and uses an angle of 45 degrees to the horizontal. **43**

Orthographic drawing Formal working drawing that shows an object from every angle, to help manufacturers plan production; it uses several 2D views to represent a 3D object. **44**

Overheads Costs or expenses, such as rent, insurance, lighting, heating and equipment, that are paid out by an organisation. **11, 35**

Planned obsolescence Policy of producing consumer goods that rapidly become out of date and unusable, either due to changes in design or unavailability of spare parts. **38**

Press fit Mechanical, solderless method of tube joining. **33**

Press forming Forming operation used to change the shape of a sheet material using pressure from a press. **33**

Product life cycle The various stages of a product's evolution, from its beginning to its end. **10**

Prosecuted Officially accused in a court of breaking the law. **38**

Prototype To create a 3D model that demonstrates the functionality of a product. **11**

Qualitative Cannot be expressed as a number and describes opinions, qualities and feelings. **23**

Quality function deployment (QFD) Evaluation tool that helps transform the customer's needs and wants into a product design. **63**

Quantitative Can be expressed as a number and quantified by hard facts. **23**

Radius A straight line from the centre of a circle to its circumference (outer edge). **52**

Ranking matrices Tables with numbers to support decision making when comparing products when there is more than one option and several factors to consider. **61**

Rivets Mechanical fasteners with a head on one end and a cylindrical stem on the other. **33**

Scale Amount by which a drawing is enlarged or reduced from the actual size of an object, shown as a ratio. **50**

Shaping Process that involves a change in state of a material. **30**

Shears Tool that cuts material with a scissor action by applying opposing forces on opposite sides, forcing the material apart. **29**

Sintering Process that turns metal powders into solid objects using high temperatures. **30**

Soldering Joining different types of metal by melting solder. **33**

Specification A list of criteria that a product needs to address. **17**

Standards Guidelines on how to meet legislation that set out agreed ways of doing something, such as making a product, managing a process or supplying materials. **37**

STL Abbreviation for a stereolithography file used in CAD software systems to create 3D objects. **67**

Stock form The shape in which materials are available. **28**

Subjective Based on personal feelings, tastes or opinions. **23**

Supply and demand Relationship between the quantity of products a business has available to sell and the amount consumers want to buy. **28**

Sustainable Using something in a way that ensures it does not run out and affect the needs of future generations. **39**

Sustainable design Development of a product while trying to reduce negative impacts on the environment. **14**

Target market The group of people a product is made for. **14**

Technology push When new technology is created as a result of research and development (R&D), resulting in new products that are 'pushed' into the market, with or without demand. **37**

Thermoplastic polymers Polymers that become pliable when heated, such that they can be shaped and formed, and harden when cooled; this process can be repeated over and over again. **30**

Thermosetting polymers Polymers that can be shaped and formed when heated, but this process can only occur once; they cannot be reheated or reformed. **30**

Thread Spiral grooves of equal measurement on a cylindrical article or pipe; threads can be on the inside (nut) or outside (screw) of a cylinder. **29, 56**

Threaded fasteners Products such as screws, nuts and bolts. **33**

Through holes Holes that break through to the other side of the workpiece. **56**

Tolerance Amount of variation allowed in a dimension if the product is to work as intended. **34**

User-centred design Development of a product with a clear understanding of user needs and requirements. **14**

Users People who will use the final product. **10**

Virtual modelling Producing a replica of a product or component using a software package, which can be tested without the need to make a physical model. **65**

Want Non-essential but desirable aspect of a product. **23**

Wasting Process that removes material. **29**

Welding Joining metal parts where the part edges are melted by means of heat. **33**

Wiring diagrams Simplified pictorial representations of circuits that show how each component should be connected. **47**